WHOLE NUMBERS
Addition & Subtraction

Allan D. Suter

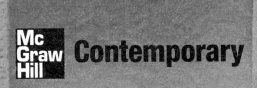

McGraw Hill Contemporary

Series Editor: Mitch Rosin
Executive Editor: Linda Kwil
Production Manager: Genevieve Kelley
Marketing Manager: Sean Klunder
Cover Design: Steve Strauss, ¡Think! Design

Send all inquiries to:
McGraw-Hill/Contemporary
130 East Randolph Street, Suite 400
Chicago, Illinois 60601

ISBN: 0-07-287104-0

Printed in the United States of America.

1 2 3 4 5 6 7 8 9 10 QPD/QPD 08 07 06 05 04 03

The **McGraw·Hill** Companies

■ Contents

1.
$$58 + 49$$

Answer: _____

6.
$$437 \\ 25 \\ +619$$

Answer: _____

2. $5{,}209 + 56 + 164 =$

Answer: _____

7. $36 + 82 + 17 =$

Answer: _____

3.
$$649 \\ 935 \\ + \ 85$$

Answer: _____

8.
$$93 \\ - \ 9$$

Answer: _____

4. $36 + 6{,}409 + 767 =$

Answer: _____

9. $50 - 24 =$

Answer: _____

5. $73 + 48 =$

Answer: _____

10. $705 - 136 =$

Answer: _____

11.
$$
\begin{array}{r}
837 \\
-458 \\
\hline
\end{array}
$$

Answer: _____

12.
$$
\begin{array}{r}
63 \\
-29 \\
\hline
\end{array}
$$

Answer: _____

13.
$$
\begin{array}{r}
306 \\
-198 \\
\hline
\end{array}
$$

Answer: _____

14. $8,102 - 483 =$

Answer: _____

15. Don deposited $75 into his savings account. He now has $350 in the account. How much did he have before he deposited the $75?

Answer: _____

16. John's monthly rent of $485 was increased $35. How much is the monthly rent after the increase?

Answer: _____

17. Lin sold her car for $3,595. She bought the car four years earlier for $6,200. The price she sold the car for was how much less than the price she paid?

Answer: _____

18. For lunch Richard ate a slice of pizza that had 565 calories and a root beer with 85 calories. Altogether how many calories were in his lunch?

Answer: _____

19. Wednesday morning Kaye drove 117 miles to Central City where she stopped for lunch. Then she drove 79 miles farther to Greenport. What total distance did she drive that day?

Answer: _____

20. Bill's new car needs a tune-up once he has driven 20,000 miles. The odometer on his car says that he has driven 18,309 miles. How many more miles should he drive before he gets a tune-up?

Answer: _____

Evaluation Chart

On the following chart, circle the number of any problem you missed. The column after the problem number tells you the pages where those problems are taught. Based on your score, your teacher may ask you to study specific sections of this book. However, to thoroughly review your skills, begin with Unit 1 on page 7.

Skill Area	Pretest Problem Number	Skill Section	Review Page
Addition	1, 2, 3, 4, 5, 6, 7	17–31	32
Subtraction	8, 9, 10, 11, 12, 13, 14	39–60	61
Addition Problem Solving	16, 18, 19	33–37	38
Subtraction Problem Solving	15, 17, 20	63–68	69
Life-Skills Math	All	70–73	74

Hundreds

To understand and work with numbers, you need to learn about place value. Let's start with ones, tens, and hundreds.

Fill in the blanks at the right.

1. ⎢ ⎢8⎢9⎢ Eighty-nine has

    a) __8__ tens

    b) ____ ones

2. ⎢ ⎢7⎢2⎢ Seventy-two has

    a) ____ tens

    b) ____ ones

3. ⎢4⎢0⎢5⎢ Four hundred five has

    a) ____ hundreds

    b) ____ tens

    c) ____ ones

4. ⎢9⎢1⎢3⎢ Nine hundred thirteen has

    a) ____ hundreds

    b) ____ tens

    c) ____ ones

5. ⎢3⎢0⎢0⎢ Three hundred has

    a) ____ hundreds

    b) ____ tens

    c) ____ ones

Thousands

Numbers in the thousands have more than 3 digits. A number in the thousands has a comma to separate the thousands from the hundreds.

thirty-nine thousand   39 , 516   five hundred sixteen

Write the correct digit in each place-value position.

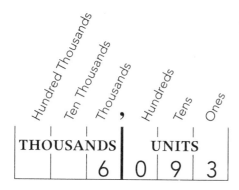

1. Six thousand, ninety-three

     a) __6__ thousands

     b) ____ hundreds

     c) ____ tens

     d) ____ ones

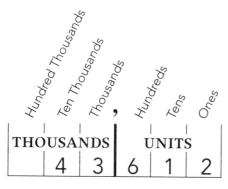

2. Forty-three thousand, six hundred twelve

     a) ____ ten thousands

     b) ____ thousands

     c) ____ hundreds

     d) ____ tens

     e) ____ ones

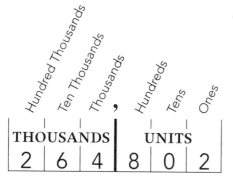

3. Two hundred sixty-four thousand, eight hundred two

     a) ____ hundred thousands

     b) ____ ten thousands

     c) ____ thousands

     d) ____ hundreds

     e) ____ tens

     f) ____ ones

Place-Value Readiness

Print the words from the list on the place-value chart.

LIST
hundreds
ones
ten thousands
tens
thousands

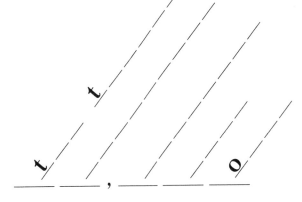

PLACE-VALUE CHART

Use 23,547 for problems 1–5.

1. The 3 has what place value? t h ___ ___ ___ ___ ___ ___ ___ ___

2. The 4 has what place value? ___ ___ ___ ___

3. The 5 has what place value? ___ ___ ___ ___ ___ ___ ___ ___

4. The 2 has what place value? t ___ ___ t ___ ___ ___ ___ ___ ___ ___ ___

5. The 7 has what place value? ___ ___ ___ ___

Use 87,325 for problems 6–10.

6. What digit is in the **ones** place? ___

7. What digit is in the **hundreds** place? ___

8. What digit is in the **ten thousands** place? ___

9. What digit is in the **thousands** place? ___

10. What digit is in the **tens** place? ___

Number Sense

Sometimes we need to use numbers that are larger than thousands. We may talk about millions of dollars or billions of stars.

47,000,000
forty-seven million

3,000,000,000
three billion

Place-Value Chart

Fill in and line up place values.

	Hundred Billions	Ten Billions	Billions	Hundred Millions	Ten Millions	Millions	Hundred Thousands	Ten Thousands	Thousands	Hundreds	Tens	Ones
	BILLIONS			**MILLIONS**			**THOUSANDS**			**UNITS**		
1. 689 →										6	8	9
2. 6,093 →												
3. 149,508 →												
4. 23,154 →												
5. 5,904,165 →												
6. 196 →												
7. 62,949,150 →												
8. 3,992 →												
9. 649,132,001 →												
10. 326,194,600,109 →												

Commas

Large numbers are hard to read without commas to separate the hundreds, thousands, millions, and billions.

Copy the numbers below in the spaces.

A. 346   __ __ 6

B. 12,346   __ __ , __ __ 6

C. 21,612,346   __ __ , __ __ __ , __ __ 6

D. 7,421,612,346   __ , __ __ __ , __ __ __ , __ __ 6

Start at the right and count 3 places.
Every 3 places, write a comma.

Write the commas where they are needed.

1. 7349 =   7 , 3 4 9
                       3 2 1

7. 1 4 1 3 9 6 2

2. 21496 =   __ __ , __ __ __

8. 2 1 7 9 2 4

3. 14948 =   __ __ __ __ __

9. 1 4 3 9

4. 964821 =   __ __ __ __ __ __
               ↑—no comma in front

10. 3 4 1

5. 1436849 =   __ __ __ __ __ __ __

11. 1 1 6 8 2 3 7 1

6. 29743268 = __ __ __ __ __ __ __ __

12. 4 6 2 1 8

Words to Numbers

Remember that commas are placed between every three digits of a number—moving from right to left.

Write the number for each.

1. Five hundred nine   <u>5</u> <u>0</u> <u>9</u>

             ↑——— use a zero to hold this place

2. Seven hundred three   __ __ __

3. Two thousand, one hundred sixteen   __ , __ __ __

4. Five thousand, eight   __ , __ __ __

5. Six thousand, eight hundred forty-three   __ , __ __ __

6. Seventy-four thousand, nine hundred four   __ __ , __ __ __

7. Seven thousand, four   __ , __ __ __

8. Three hundred seventeen million, fifty-two thousand

   ↑ __ __ __ , __ __ __ , __ __ __

      └——— no comma in front

9. Twenty-nine billion, seven hundred four million, five hundred thousand, three hundred ninety-six

   __ __ , __ __ __ , __ __ __ , __ __ __

Write the numbers, and be sure to put in commas.

10. Nine thousand, fifteen   _____

11. Twenty thousand, forty-five   _____

12. Thirty million, four hundred thirty-six thousand, one hundred sixty-five

   _____

Expanded Forms

Expanded forms of numbers can help you understand place value.

<u>EXAMPLE 1</u>

$$689 = (6 \times 100) + (8 \times 10) + (9 \times 1)$$
$$= \quad 600 \quad + \quad 80 \quad + \quad 9$$

<u>EXAMPLE 2</u>

$$23{,}154 = (2 \times 10{,}000) + (3 \times 1{,}000) + (1 \times 100) + (5 \times 10) + (4 \times 1)$$
$$= \quad 20{,}000 \quad + \quad 3{,}000 \quad + \quad 100 \quad + \quad 50 \quad + \quad 4$$

Complete the expanded forms.

1. $79 = (\underline{} \times 10) + (\underline{} \times 1)$

$\quad = \quad 70 \quad + \quad 9$

2. $196 = (\underline{} \times 100) + (\underline{} \times \underline{}) + (\underline{} \times \underline{})$

$\quad = \underline{} + \underline{} + \underline{}$

3. $6{,}093 = (\underline{} \times 1{,}000) + (0 \times \underline{}) + (\underline{} \times 10) + (\underline{} \times 1)$

$\quad = \underline{} + \underline{} + \underline{} + \underline{}$

4. $4{,}980 = (\underline{} \times 1{,}000) + (\underline{} \times 100) + (\underline{} \times \underline{}) + (\underline{} \times \underline{})$

$\quad = \underline{} + \underline{} + \underline{} + \underline{}$

5. $5{,}146 = (\underline{} \times \underline{}) + (\underline{} \times \underline{}) + (\underline{} \times \underline{}) + (\underline{} \times \underline{})$

$\quad = \underline{} + \underline{} + \underline{} + \underline{}$

Ordering Numbers

Arrange the numbers from **smallest** to **largest**.

83   651   115   380   4,022

1. _____ _____ _____ _____ _____
  smallest   largest

Arrange the numbers from **largest** to **smallest**.

75   290   136   802   1,094

2. _____ _____ _____ _____ _____
  largest   smallest

Rearrange the 3 digits to name the **smallest** possible number.

3. 581   _1_ _5_ _8_

4. 723   ___ ___ ___

5. 492   ___ ___ ___

Rearrange the 4 digits to name the **largest** possible number.

6. 2704 ___,___ ___ ___

7. 3825 ___,___ ___ ___

8. 9061 ___,___ ___ ___

Rearrange the 5 digits to name the **largest** possible number.

9. 43075 ___ ___,___ ___ ___

10. 39154 ___ ___,___ ___ ___

Comparing Numbers

NUMBER RELATION SYMBOLS		
SYMBOL	MEANING	EXAMPLES
=	is equal to	5 = 5 5 is equal to 5
<	is less than	7 < 8 7 is less than 8
>	is greater than	6 > 3 6 is greater than 3

Draw a circle around the number that makes each a true statement.

1. 48 = _____   84   48   38

2. 315 < _____   315   135   513
    is less than

3. 9,408 > _____   9,804   9,408   9,048
    is greater than

4. 52,546 < _____   52,546   55,246   25,465

Place the symbols < , = , or > in the ◯ to make each statement true.

5. 105 ◯ 110   9. 2,202 ◯ 2,020

6. 150 ◯ 130   10. 52,904 ◯ 42,509

7. 1,394 ◯ 1,349   11. 210 ◯ 120

8. 1,005 ◯ 1,005   12. 41,052 ◯ 41,042

Place Value Review

1. The four has what place value?

   32,498

   _____

2. The seven has what place value?

   742,198

   _____

3. What digit is in the thousands place?

   76,387

   _____

4. Write the commas where they are needed.

   2 3 9 4 7 3 8

5. Write the number.

   One hundred fifty-nine thousand, six hundred forty-two

   _____

6. Write the expanded form.

   4,826 = (___ × 1,000) + (___ × 100) + (___ × 10) + (___ × 1)

7. Rearrange the digits to name the smallest possible number.

   9 4 2

   _____

8. Rearrange the digits to name the largest possible number.

   5 9 2 6 4

   _____

9. Place the symbols <, =, or > in the ◯ to make the statement true.

   230 ◯ 320

10. Place the symbols <, =, or > in the ◯ to make the statement true.

    564 ◯ 645

Meaning of Addition

Addition means "to continue" or "to join together." A "plus" sign $\boxed{+}$ is the symbol for addition.

Use the picture to fill in the blanks.

A. _____ + _____ = _____
     shoes   shoes   shoes in all

B.
$$\begin{array}{r} 2 \\ +\ 4 \\ \hline \boxed{} \end{array}$$

Complete each number sentence.

1.

_____ + _____ = _____
circles   circles circles in all

3. Draw 3 circles.   Draw 5 circles.

_____ + _____ = _____
circles   circles circles in all

2.

_____ + _____ = _____
circles   circles circles in all

4. Draw 5 circles.   Draw 8 circles.

_____ + _____ = _____
circles   circles circles in all

Addition Facts

To add and subtract, you must memorize the addition facts. Add each number in the rows (going across) to the numbers in the columns (going down).

Follow the example of 2 + 3 = 5.

+	+0	+1	+2	+3
1				
2				5
3				

Fill in the chart. Go across each row. Start with 1 + 0 = 1.

+	+0	+1	+2	+3	+4	+5	+6	+7	+8	+9
1	1	2								
2			4							
3				6						
4					8					
5						10				
6							12			
7								14		
8									16	
9										18

Practice Helps

Each column should be done in less than one minute.

1. $9 + 5 =$ ___	21. $8 + 9 =$ ___	41. $3 + 8 =$ ___	61. $6 + 6 =$ ___
2. $3 + 8 =$ ___	22. $9 + 7 =$ ___	42. $8 + 8 =$ ___	62. $6 + 9 =$ ___
3. $7 + 8 =$ ___	23. $3 + 7 =$ ___	43. $7 + 7 =$ ___	63. $9 + 7 =$ ___
4. $5 + 6 =$ ___	24. $6 + 9 =$ ___	44. $5 + 7 =$ ___	64. $6 + 8 =$ ___
5. $4 + 9 =$ ___	25. $9 + 9 =$ ___	45. $6 + 6 =$ ___	65. $9 + 4 =$ ___
6. $7 + 9 =$ ___	26. $9 + 8 =$ ___	46. $9 + 6 =$ ___	66. $6 + 5 =$ ___
7. $9 + 6 =$ ___	27. $5 + 8 =$ ___	47. $9 + 9 =$ ___	67. $8 + 7 =$ ___
8. $9 + 4 =$ ___	28. $8 + 8 =$ ___	48. $9 + 4 =$ ___	68. $6 + 7 =$ ___
9. $9 + 3 =$ ___	29. $9 + 3 =$ ___	49. $6 + 5 =$ ___	69. $5 + 8 =$ ___
10. $8 + 8 =$ ___	30. $8 + 9 =$ ___	50. $7 + 8 =$ ___	70. $3 + 9 =$ ___
11. $8 + 5 =$ ___	31. $8 + 6 =$ ___	51. $8 + 3 =$ ___	71. $8 + 6 =$ ___
12. $9 + 8 =$ ___	32. $7 + 7 =$ ___	52. $5 + 9 =$ ___	72. $7 + 7 =$ ___
13. $6 + 7 =$ ___	33. $7 + 5 =$ ___	53. $4 + 8 =$ ___	73. $4 + 7 =$ ___
14. $6 + 0 =$ ___	34. $6 + 6 =$ ___	54. $9 + 8 =$ ___	74. $0 + 4 =$ ___
15. $6 + 8 =$ ___	35. $9 + 6 =$ ___	55. $9 + 0 =$ ___	75. $4 + 5 =$ ___
16. $6 + 9 =$ ___	36. $7 + 0 =$ ___	56. $4 + 4 =$ ___	76. $6 + 4 =$ ___
17. $7 + 6 =$ ___	37. $7 + 9 =$ ___	57. $4 + 6 =$ ___	77. $7 + 5 =$ ___
18. $3 + 7 =$ ___	38. $4 + 9 =$ ___	58. $7 + 3 =$ ___	78. $5 + 5 =$ ___
19. $7 + 4 =$ ___	39. $6 + 5 =$ ___	59. $7 + 4 =$ ___	79. $7 + 9 =$ ___
20. $9 + 7 =$ ___	40. $8 + 7 =$ ___	60. $7 + 5 =$ ___	80. $9 + 5 =$ ___

Counting Patterns

1. Count by 5 starting at 5.

   <u>5</u> <u>10</u> __ __ __ __ __ __ __ __ __ __ <u>60</u>

2. Count by 10 starting at 10.

   <u>10</u> <u>20</u> __ __ __ __ __ __ __ __ __ __ <u>120</u>

3. Count by 2 starting at 3.

   <u>3</u> <u>5</u> __ __ __ __ __ __ __ __ __ __ <u>25</u>

4. Count by 6 starting at 13.

   <u>13</u> <u>19</u> __ __ __ __ __ __ __ __ __ __ <u>79</u>

5. Count by 4 starting at 5.

   <u>5</u> <u>9</u> __ __ __ __ __ __ __ __ __ __ <u>49</u>

6. Count by 3 starting at 0.

   <u>0</u> <u>3</u> __ __ __ __ __ __ __ __ __ __ <u>33</u>

Continue the patterns.

7. +7 +7
    <u>3</u> <u>10</u> <u>17</u> __ __ __ __ __ <u>59</u>

8. +□ +□
    <u>4</u> <u>13</u> <u>22</u> __ __ __ __ __ <u>76</u>

9. +□ +□
    <u>9</u> <u>18</u> <u>27</u> __ __ __ __ __ <u>81</u>

10. <u>7</u> <u>15</u> <u>23</u> __ __ __ __ __ <u>71</u>

11. <u>11</u> <u>17</u> <u>23</u> __ __ __ __ __ <u>59</u>

12. <u>9</u> <u>16</u> <u>23</u> __ __ __ __ __ <u>65</u>

Timed Addition Drill

Can you add these in less than 3 minutes?

1. $\begin{array}{r} 6 \\ +\ 8 \\ \hline \end{array}$	10. $\begin{array}{r} 7 \\ +\ 3 \\ \hline \end{array}$	19. $\begin{array}{r} 3 \\ +\ 4 \\ \hline \end{array}$
2. $\begin{array}{r} 10 \\ +\ 8 \\ \hline \end{array}$	11. $\begin{array}{r} 5 \\ +\ 7 \\ \hline \end{array}$	20. $\begin{array}{r} 7 \\ +\ 4 \\ \hline \end{array}$
3. $\begin{array}{r} 10 \\ +10 \\ \hline \end{array}$	12. $\begin{array}{r} 7 \\ +\ 8 \\ \hline \end{array}$	21. $\begin{array}{r} 5 \\ +\ 4 \\ \hline \end{array}$
4. $\begin{array}{r} 9 \\ +10 \\ \hline \end{array}$	13. $\begin{array}{r} 4 \\ +\ 7 \\ \hline \end{array}$	22. $\begin{array}{r} 3 \\ +\ 3 \\ \hline \end{array}$
5. $\begin{array}{r} 4 \\ +\ 5 \\ \hline \end{array}$	14. $\begin{array}{r} 9 \\ +\ 5 \\ \hline \end{array}$	23. $\begin{array}{r} 6 \\ +\ 3 \\ \hline \end{array}$
6. $\begin{array}{r} 5 \\ +\ 5 \\ \hline \end{array}$	15. $\begin{array}{r} 9 \\ +\ 7 \\ \hline \end{array}$	24. $\begin{array}{r} 2 \\ +\ 7 \\ \hline \end{array}$
7. $\begin{array}{r} 8 \\ +\ 2 \\ \hline \end{array}$	16. $\begin{array}{r} 6 \\ +10 \\ \hline \end{array}$	25. $\begin{array}{r} 9 \\ +\ 8 \\ \hline \end{array}$
8. $\begin{array}{r} 8 \\ +\ 6 \\ \hline \end{array}$	17. $\begin{array}{r} 7 \\ +\ 6 \\ \hline \end{array}$	26. $\begin{array}{r} 4 \\ +\ 4 \\ \hline \end{array}$
9. $\begin{array}{r} 4 \\ +\ 6 \\ \hline \end{array}$	18. $\begin{array}{r} 10 \\ +\ 7 \\ \hline \end{array}$	27. $\begin{array}{r} 8 \\ +\ 7 \\ \hline \end{array}$

28. $\begin{array}{r} 5 \\ +\ 9 \\ \hline \end{array}$	37. $\begin{array}{r} 7 \\ +\ 9 \\ \hline \end{array}$	46. $\begin{array}{r} 8 \\ +\ 4 \\ \hline \end{array}$
29. $\begin{array}{r} 10 \\ +\ 9 \\ \hline \end{array}$	38. $\begin{array}{r} 3 \\ +\ 9 \\ \hline \end{array}$	47. $\begin{array}{r} 7 \\ +10 \\ \hline \end{array}$
30. $\begin{array}{r} 6 \\ +\ 5 \\ \hline \end{array}$	39. $\begin{array}{r} 9 \\ +\ 9 \\ \hline \end{array}$	48. $\begin{array}{r} 6 \\ +\ 6 \\ \hline \end{array}$
31. $\begin{array}{r} 9 \\ +\ 2 \\ \hline \end{array}$	40. $\begin{array}{r} 7 \\ +\ 1 \\ \hline \end{array}$	49. $\begin{array}{r} 8 \\ +\ 8 \\ \hline \end{array}$
32. $\begin{array}{r} 10 \\ +\ 3 \\ \hline \end{array}$	41. $\begin{array}{r} 2 \\ +\ 5 \\ \hline \end{array}$	50. $\begin{array}{r} 8 \\ +10 \\ \hline \end{array}$
33. $\begin{array}{r} 8 \\ +\ 3 \\ \hline \end{array}$	42. $\begin{array}{r} 0 \\ +\ 6 \\ \hline \end{array}$	51. $\begin{array}{r} 2 \\ +\ 3 \\ \hline \end{array}$
34. $\begin{array}{r} 3 \\ +\ 5 \\ \hline \end{array}$	43. $\begin{array}{r} 6 \\ +\ 9 \\ \hline \end{array}$	52. $\begin{array}{r} 3 \\ +\ 6 \\ \hline \end{array}$
35. $\begin{array}{r} 4 \\ +\ 8 \\ \hline \end{array}$	44. $\begin{array}{r} 9 \\ +\ 4 \\ \hline \end{array}$	53. $\begin{array}{r} 5 \\ +10 \\ \hline \end{array}$
36. $\begin{array}{r} 4 \\ +10 \\ \hline \end{array}$	45. $\begin{array}{r} 9 \\ +\ 3 \\ \hline \end{array}$	54. $\begin{array}{r} 4 \\ +\ 3 \\ \hline \end{array}$

Column Addition

When you add 3 or more numbers, you only add 2 numbers at a time. Complete the column problems by first adding 2 numbers at a time.

1.
$$\begin{array}{r} 6 \\ 8 \\ +\ 7 \\ \hline 21 \end{array}$$
think
$14 + 7$

5.
$$\begin{array}{r} 3 \\ 5 \\ 8 \\ +\ 4 \\ \hline \end{array}$$
think
8
$+12$

9.
$$\begin{array}{r} 8 \\ 4 \\ 3 \\ +\ 2 \\ \hline \end{array}$$

2.
$$\begin{array}{r} 9 \\ 5 \\ +\ 6 \\ \hline \end{array}$$
fill in
$\Box + 6$

6.
$$\begin{array}{r} 9 \\ 6 \\ 5 \\ +\ 3 \\ \hline \end{array}$$

10.
$$\begin{array}{r} 5 \\ 4 \\ 1 \\ +\ 9 \\ \hline \end{array}$$

3.
$$\begin{array}{r} 4 \\ 5 \\ +\ 8 \\ \hline \end{array}$$
fill in
$\Box + \Box$

7.
$$\begin{array}{r} 8 \\ 7 \\ 3 \\ +\ 6 \\ \hline \end{array}$$

11.
$$\begin{array}{r} 3 \\ 6 \\ 7 \\ +\ 5 \\ \hline \end{array}$$

4.
$$\begin{array}{r} 2 \\ 7 \\ +\ 5 \\ \hline \end{array}$$
fill in
$\Box + \Box$

8.
$$\begin{array}{r} 8 \\ 3 \\ 7 \\ +\ 8 \\ \hline \end{array}$$

12.
$$\begin{array}{r} 7 \\ 8 \\ 7 \\ 6 \\ +\ 7 \\ \hline \end{array}$$
15
$\Box$ fill in
$\rightarrow\ +7$

Two-Digit Addition

When adding two-digit numbers, always start at the ones place and work left.

- Add the ones.

- Add the tens.

```
    36          A.   54        B.   73        C.   46
  + 52             + 41           + 10           + 32
  ────            ────           ────           ────
    88                 5              3              8
```
↑ start here   finish   finish   finish

Add the numbers.

1.   15
    + 24
    ────

2.   31
    + 22
    ────

3.   50
    + 37
    ────

4.   82
    + 17
    ────

5.   10
    + 89
    ────

6.   13
    + 35
    ────

7.   37
    + 11
    ────

8.   40
    + 20
    ────

9.   28
    + 70
    ────

10.   17
     + 72
     ────

11.   44
     + 15
     ────

12.   32
     + 45
     ────

13.   33
     + 34
     ────

14.   66
     + 11
     ────

15.   36
     + 52
     ────

16.   55
     + 23
     ────

Three-Digit Addition

Remember to start at the ones place and work to the left.

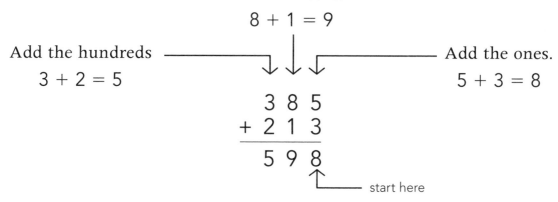

Add the tens.
8 + 1 = 9

Add the hundreds
3 + 2 = 5

Add the ones.
5 + 3 = 8

```
  3 8 5
+ 2 1 3
─────────
  5 9 8
```
↑ start here

Add the three-digit numbers.

1.
```
  765
+214
```

5.
```
  176
+322
```

9.
```
  845
+102
```

13.
```
  603
+332
```

2.
```
  712
+136
```

6.
```
  358
+340
```

10.
```
  698
+201
```

14.
```
  135
+114
```

3.
```
  322
+152
```

7.
```
  331
+200
```

11.
```
  162
+211
```

15.
```
  355
+231
```

4.
```
  202
+303
```

8.
```
  273
+314
```

12.
```
  317
+532
```

16.
```
  462
+113
```

Whole Numbers: Addition & Subtraction

Find the Missing Number

Fill in the number that completes the problem.

1. $5 + \underline{2} = 7$	21. $2 + \underline{} = 5$	41. $\underline{} + 0 = 3$	61. $\underline{} + 7 = 8$
2. $3 + \underline{} = 6$	22. $4 + \underline{} = 9$	42. $\underline{} + 6 = 8$	62. $5 + \underline{} = 7$
3. $8 + \underline{} = 8$	23. $5 + \underline{} = 5$	43. $\underline{} + 2 = 7$	62. $\underline{} + 2 = 5$
4. $4 + \underline{} = 9$	24. $6 + \underline{} = 7$	44. $\underline{} + 1 = 9$	64. $7 + \underline{} = 8$
5. $0 + \underline{} = 3$	25. $1 + \underline{} = 9$	45. $\underline{} + 4 = 4$	65. $4 + \underline{} = 7$
6. $6 + \underline{} = 8$	26. $0 + \underline{} = 7$	46. $\underline{} + 3 = 5$	66. $\underline{} + 1 = 5$
7. $2 + \underline{} = 7$	27. $3 + \underline{} = 8$	47. $\underline{} + 8 = 9$	67. $\underline{} + 9 = 9$
8. $9 + \underline{} = 9$	28. $\underline{4} + 2 = 6$	48. $\underline{} + 2 = 4$	68. $9 + \underline{} = 9$
9. $5 + \underline{} = 8$	29. $\underline{} + 5 = 9$	49. $\underline{} + 4 = 9$	69. $8 + \underline{} = 11$
10. $1 + \underline{} = 6$	30. $\underline{} + 3 = 3$	50. $\underline{} + 6 = 7$	70. $4 + \underline{} = 7$
11. $4 + \underline{} = 8$	31. $\underline{} + 4 = 7$	51. $\underline{} + 4 = 8$	71. $\underline{} + 1 = 9$
12. $3 + \underline{} = 7$	32. $\underline{} + 6 = 9$	52. $\underline{} + 2 = 3$	72. $7 + \underline{} = 8$
13. $0 + \underline{} = 5$	32. $\underline{} + 1 = 5$	53. $\underline{} + 1 = 7$	73. $\underline{} + 2 = 6$
14. $2 + \underline{} = 6$	34. $\underline{} + 3 = 6$	54. $\underline{} + 7 = 9$	74. $\underline{} + 5 = 8$
15. $4 + \underline{} = 7$	35. $\underline{} + 2 = 8$	55. $\underline{} + 3 = 6$	75. $7 + \underline{} = 7$
16. $2 + \underline{} = 8$	36. $\underline{} + 0 = 7$	56. $\underline{} + 4 = 7$	76. $\underline{} + 2 = 7$
17. $3 + \underline{} = 9$	37. $\underline{} + 3 = 9$	57. $\underline{} + 3 = 3$	77. $\underline{} + 3 = 9$
18. $7 + \underline{} = 9$	38. $\underline{} + 2 = 5$	58. $\underline{} + 4 = 6$	78. $2 + \underline{} = 8$
19. $4 + \underline{} = 6$	39. $\underline{} + 7 = 9$	59. $\underline{} + 9 = 9$	79. $8 + \underline{} = 9$
20. $5 + \underline{} = 9$	40. $\underline{} + 5 = 8$	60. $\underline{} + 2 = 7$	80. $\underline{} + 0 = 3$

Regroup the Ones

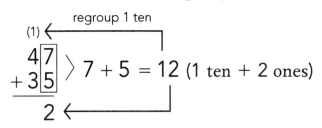

STEP 1
Add the ones and regroup.

regroup 1 ten

(1)

4 7
+ 3 5
—————
2

$7 + 5 = 12$ (1 ten + 2 ones)

STEP 2
Add the tens.

(1)

4 7
+ 3 5
—————
8 2

Add and regroup.

1. (1)
    1 6
    + 3 7
    ———
    3
    finish

2. (1)
    2 7
    + 6 8
    ———
    5
    finish

3. ( )
    3 3
    + 5 9
    ———
    2

4. ( )
    6 9
    + 6 2
    ———
    1

5. ( )
    7 4
    + 4 9
    ———

6. ( )
    5 8
    + 9 5
    ———

7. ( )
    6 2
    + 5 8
    ———

8. ( )
    5 8
    + 5 9
    ———

9. 3 5
    + 9 9
    ———

10. 8 4
    + 7 7
    ———

11. 5 3
    + 5 7
    ———

12. 2 6
    + 8 5
    ———

13. 9 8
    + 8 7
    ———

14. 9 6
    + 6 4
    ———

15. 4 8
    + 8 8
    ———

16. 3 4
    + 9 7
    ———

Regroup and Think Zero

Sometimes you will need to think zero when adding.

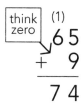

	(1)
	6 5
+	9
	7 4

	(1)
A.	5 4
+	8
	2
	finish

	( )
B.	9 8
+	7
	5
	finish

	( )
C.	7 3
+	7
	0
	finish

Add the numbers.

1. 36
  + 6

2. 92
  + 8

3. 22
  + 9

4. 13
  + 7

5. 48
  + 3

6. 79
  + 9

7. 41
  + 9

8. 77
  + 7

9. 54
  + 7

10. 63
  + 9

11. 25
  + 6

12. 67
  + 7

13. 44
  + 6

14. 66
  + 8

15. 58
  + 2

16. 66
  + 9

17. 38
  + 3

18. 78
  + 5

19. 94
  + 7

20. 19
  + 4

Adding More Numbers

Regroup 1 hundred from the tens place. ⟶

⟵ Regroup 2 tens from the ones place.

⟵ Add 4 + 8 + 9 = 21.

```
    (1)(2)
     6 9 4
     9 6 8
 +   4 2 9
 ─────────
   2,0 9 1
```

↑ ⟵ Always start at the ones place and work left.

Add the numbers.

1.
```
    ( )(1)
     8 6 9
     1 1 3
 +   4 9 6
 ─────────
 _,_ _ 8
   finish
```

3.
```
    ( )(1)
     2 8 3
     5 5 4
 +   9 3 6
 ─────────
 _,_ _ 3
   finish
```

5.
```
    ( )(1)
     3 1 4
     6 3 3
 +   4 6 5
 ─────────
 _,_ _ 2
   finish
```

2.
```
     1 9 5
     3 6 8
 +   7 4 8
 ─────────
```

4.
```
     6 4 9
     2 5 7
 +   3 8 4
 ─────────
```

6.
```
     9 3 5
     7 0 8
 +   1 2 7
 ─────────
```

Addition Practice

Add the numbers.

1. $45 + 23 =$

2. $\begin{array}{r} 84 \\ +76 \\ \hline \end{array}$

3. $34 + 42 + 93 =$

4. $\begin{array}{r} 76 \\ 25 \\ +89 \\ \hline \end{array}$

5. $353 + 526 =$

6. $\begin{array}{r} 945 \\ +178 \\ \hline \end{array}$

7. $325 + 483 + 938 =$

8. $\begin{array}{r} 3,298 \\ 7,120 \\ +4,902 \\ \hline \end{array}$

Using a Grid

Line up place values **without working the problems.** Be sure to use commas where needed.

1. 9,304 + 92 + 645 =

9	3	0	4
		9	2

5. 95 + 6 + 139 + 6,480 =

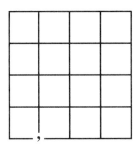

2. 32 + 195 + 6,452 =

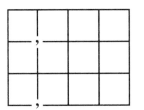

6. 8,956 + 88,309 + 6 =

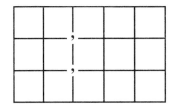

3. 8,894 + 65 + 3,205 =

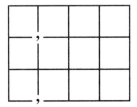

7. 23,914 + 56 + 9 + 13 =

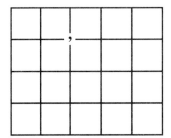

4. 39 + 2,503 + 4 =

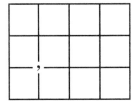

8. 392 + 8 + 94,138 + 14 + 4 =

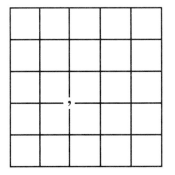

Whole Numbers: Addition & Subtraction

Lining Up Numbers

Line up the numbers and add.

1. 7,035 + 49 + 6,430 =

$$
\begin{array}{r}
7,035 \\
49 \\
+\ 6,430 \\
\hline
\text{--,---} \\
\end{array}
$$
finish

2. 535 + 29 + 7,315 =

3. 35 + 908 + 167 =

4. 5,362 + 57 + 2,901 =

5. 57 + 9 + 7,045 =

6. 5,913 + 273 + 104 =

7. 463 + 17 + 1,472 =

8. 9,631 + 173 + 85 =

Addition Review

1. For each number, circle the digit in the thousands place.

   a) 4,932 b) 609,056 c) 54,962

2. For each number, circle the digit in the hundreds place.

   a) 349   b) 9,614   c) 32,100

3. For each number, circle the digit in the ones place.

   a) 84   b) 146   c) 2,296

4. For each number, circle the digit in the tens place.

   a) 93   b) 6,840   c) 61,232

7. Write this number with commas.

   4379321

   _____

8. 34 + 142 + 21 =

9. 4,813 + 16 + 3,194 =

10. 285 + 23 + 2,904 =

5. Arrange the numbers from smallest to largest:

   47,147   75   705   431   9,103

   _____   _____   _____   _____   _____
   smallest   largest

6. Write the expanded number:

   $1,932 = (\underline{\ } \times 1,000) + (9 \times \underline{\ \ \ \ }) + (\underline{\ } \times \underline{\ \ }) + (\underline{\ } \times \underline{\ })$

   $= \underline{\ \ \ \ \ \ } + \underline{\ \ \ \ \ \ } + \underline{\ \ \ \ \ \ } + \underline{\ \ \ \ \ \ }$

Addition Problems

Addition means	—	sum
	—	to add
	—	to find the total
	—	to find how many

Complete each number sentence.

1.

If you have 3 circles and add 2, how many are there?

$\underline{\ \ 3\ \ }$	$\underline{\ \ +\ \ }$	$\underline{\ \ 2\ \ }$	=	$\underline{\ \ \ \ \ \ }$
number	operation symbol	number		answer

4.

Write an addition problem using the 8 circles.

$\underline{\ \ \ \ \ \ }$	$\underline{\ \ \ \ \ \ }$	$\underline{\ \ \ \ \ \ }$	=	$\underline{\ \ \ \ \ \ }$
number	operation symbol	number		answer

2. Set A

Set B

There are 5 circles in Set A and 3 circles in Set B. Find the total of the two sets.

$\underline{\ \ \ \ \ \ }$	$\underline{\ \ \ \ \ \ }$	$\underline{\ \ \ \ \ \ }$	=	$\underline{\ \ \ \ \ \ }$
number	operation symbol	number		answer

5. Set A

Set B

Write a number sentence using Set A and Set B.

$\underline{\ \ \ \ \ \ }$	$\underline{\ \ \ \ \ \ }$	$\underline{\ \ \ \ \ \ }$	=	$\underline{\ \ \ \ \ \ }$
number	operation symbol	number		answer

3. Set A

Set B

If there are 6 circles in Set A and 4 circles in Set B, how many circles are there in all?

$\underline{\ \ \ \ \ \ }$	$\underline{\ \ \ \ \ \ }$	$\underline{\ \ \ \ \ \ }$	=	$\underline{\ \ \ \ \ \ }$
number	operation symbol	number		answer

6. Set A has 13 circles. Set B has 8 circles. Add the two sets.

$\underline{\ \ \ \ \ \ }$	$\underline{\ \ \ \ \ \ }$	$\underline{\ \ \ \ \ \ }$	=	$\underline{\ \ \ \ \ \ }$
number	operation symbol	number		answer

Problem-Solving Strategies

1. Read over the problem several times to make sure you understand it.

2. Think about the facts in the problem and what you are being asked to find.

3. Complete the number sentence for each problem.

4. Ask yourself, "Does the answer make sense?"

Addition means "to combine" or "to join together."
Clue words can help you decide to add.

added to	in all	combined	plus
total	altogether	increased	sum

Complete each number sentence.

1. $9 added to $5 is how much money?

9	+	5	=	$14
number	operation symbol	number		answer

5. 8 increased by 9 is what number?

___	___	___	=	___
number	operation symbol	number		answer

2. $17 and $15 total how many dollars?

___	+	___	=	___
number	operation symbol	number		answer

6. 16 and 9 equal how many in all?

___	___	___	=	___
number	operation symbol	number		answer

3. 19 combined with 8 is what number?

___	___	___	=	___
number	operation symbol	number		answer

7. Find the sum of 6 and 3.

___	___	___	=	___
number	operation symbol	number		answer

4. 13 plus 7 equal what number?

___	___	___	=	___
number	operation symbol	number		answer

8. $4 and $3 equal how many dollars?

___	___	___	=	___
number	operation symbol	number		answer

Addition Word Problems

Complete each number sentence.

1. Ping rode her bike 128 miles last week and 75 miles this week. How many miles did she ride in all?

$$\underline{\quad 128 \quad} + \underline{\quad 75 \quad} = \underline{\qquad}$$
number   operation symbol   number   answer

Ping rode her bike _____ miles in all.

2. Daryl wants to buy a new car for $19,954. He wants to include the extra options at a cost of $2,129. How much would he pay for the new car and options?

$$\underline{\qquad} \ \underline{\qquad} \ \underline{\qquad} = \underline{\qquad}$$
number   operation symbol   number   answer

Daryl would pay $_____ for the new car and options.

3. After election results were counted, there were 6,495 "yes" votes and 5,092 "no" votes. How many people voted altogether?

$$\underline{\qquad} \ \underline{\qquad} \ \underline{\qquad} = \underline{\qquad}$$
number   operation symbol   number   answer

_____ voted altogether.

4. The Wu family traveled 486 miles the first day and 294 miles the second day. How many miles did they travel in all?

$$\underline{\qquad} \ \underline{\qquad} \ \underline{\qquad} = \underline{\qquad}$$
number   operation symbol   number   answer

The Wu family traveled _____ miles in all.

5. There are 24,182 general admission seats and 6,759 reserved seats. How many seats are there in all?

$$\underline{\qquad} \ \underline{\qquad} \ \underline{\qquad} = \underline{\qquad}$$
number   operation symbol   number   answer

There are _____ seats in all.

6. The attendance for the first game was 48,560 and 25,709 for the second game. What was the total attendance for the two games?

$$\underline{\qquad} \ \underline{\qquad} \ \underline{\qquad} = \underline{\qquad}$$
number   operation symbol   number   answer

The total attendance for the two games was _____ .

Think About the Facts

Eric saved $35 from his paycheck.
The next week he saved $25.

Question: <u>How much money</u>

<u>did he save in all?</u>

$35	+	$25	=	$60
number	operation symbol	number		answer

Write a question and complete a number sentence for each problem. Try using all the addition words you know.

1. Rusty bought 3 CDs on Monday. On Saturday he bought 8 CDs.

Question: _____

_____

____	____	____	=	____
number	operation symbol	number		answer

2. Nicole has $73. Her paycheck was $52.

Question: _____

_____

____	____	____	=	____
number	operation symbol	number		answer

3. Terry drove 120 miles to Seattle. Then she drove 140 miles to Abbotsford.

Question: _____

_____

____	____	____	=	____
number	operation symbol	number		answer

4. Doris bought 144 paper cups for a party. Her husband bought 72.

Question: _____

_____

____	____	____	=	____
number	operation symbol	number		answer

5. Shannon has 57 pairs of earrings. She inherited another 38 pairs.

Question: _____

_____

____	____	____	=	____
number	operation symbol	number		answer

6. Mike earned $855. His wife earned $875.

Question: _____

_____

____	____	____	=	____
number	operation symbol	number		answer

Using Symbols

NUMBER RELATION SYMBOLS
$<$ less than
$>$ greater than
$=$ equal to
$\neq$ not equal to

Complete each number sentence to make it true.

1. Max saved $37 from mowing lawns and $52 from shoveling snow.

   a) How much did Max save
   mowing and shoveling?

   $\underset{\text{mowing}}{37} + \underset{\text{shoveling}}{\rule{2cm}{0.4pt}} = \underset{\text{total}}{89}$

   b) Max saved more money
   shoveling snow.

   $\underset{\text{shoveling}}{\rule{2cm}{0.4pt}} > \underset{\text{mowing}}{\rule{2cm}{0.4pt}}$

   c) Max saved less money
   mowing lawns.

   $\underset{\text{mowing}}{\rule{2cm}{0.4pt}} < \underset{\text{shoveling}}{\rule{2cm}{0.4pt}}$

   d) Max did not make the same
   amount at each job.

   $\underset{\text{mowing}}{37} \quad \underset{\text{symbol}}{\rule{2cm}{0.4pt}} \quad \underset{\text{shoveling}}{52}$

2. Joan was paid $75 on Monday and $80 on Tuesday.
   Jeremy was paid $50 on Monday and $100 on Tuesday.

   a) How much did Joan earn
   on Monday and Tuesday?

   $\underset{\text{Monday}}{75} + \underset{\text{Tuesday}}{\rule{2cm}{0.4pt}} = \underset{\text{total}}{155}$

   b) How much did Jeremy earn
   on Monday and Tuesday?

   $\underset{\text{Monday}}{\rule{2cm}{0.4pt}} + \underset{\text{Tuesday}}{100} = \underset{\text{total}}{150}$

   c) On Monday Joan made more
   than Jeremy.

   $\underset{\text{Joan}}{\rule{2cm}{0.4pt}} > \underset{\text{Jeremy}}{\rule{2cm}{0.4pt}}$

   d) On Tuesday Joan made less
   than Jeremy.

   $\underset{\text{Joan}}{\rule{2cm}{0.4pt}} < \underset{\text{Jeremy}}{\rule{2cm}{0.4pt}}$

   e) Joan and Jeremy did not make
   the same amount.

   $\underset{\text{Joan}}{155} \quad \underset{\text{symbol}}{\rule{2cm}{0.4pt}} \quad \underset{\text{Jeremy}}{150}$

Word Problem Review

Write a number sentence for each problem.

1. On the first day of registration 212 students signed up for art classes. On the second day 59 students registered. How many students registered for art classes?

   _____ _____ _____ = _____
   number operation number   answer
           symbol

2. A new shirt cost $18. A matching skirt is on sale for $32. How much do they cost altogether?

   _____ _____ _____ = _____
   number operation number   answer
           symbol

3. The gallery sold 87 tickets for the art show. An additional 15 people bought tickets at the door. How many people attended the show?

   _____ _____ _____ = _____
   number operation number   answer
           symbol

4. At the grocery store Linda spent $12 on steak and $18 dollars on other items. How much did Linda spend in all?

   _____ _____ _____ = _____
   number operation number   answer
           symbol

5. Doug watched his favorite movie 117 times. His friend John watched the same movie 89 times. How many times did they both watch the movie?

   _____ _____ _____ = _____
   number operation number   answer
           symbol

6. On the school trip, one class picked 314 strawberries. Another class picked 189 blueberries. How many berries were picked altogether?

   _____ _____ _____ = _____
   number operation number   answer
           symbol

7. There are 50 bulbs in the garden. Another 75 bulbs have been planted. How many bulbs are planted in total?

   _____ _____ _____ = _____
   number operation number   answer
           symbol

8. A used car cost $4,569. The engine repairs will cost $557. How much will it cost to buy and fix the car?

   _____ _____ _____ = _____
   number operation number   answer
           symbol

9. An airline ticket to Greece cost John $550. The hotel in Athens cost $749. How much did the ticket and hotel cost together?

   _____ _____ _____ = _____
   number operation number   answer
           symbol

10. Bobbi bought a photocopier for $139 and a printer for $99. How much did she spend in all?

   _____ _____ _____ = _____
   number operation number   answer
           symbol

Whole Numbers: Addition & Subtraction

Meaning of Subtraction

Subtraction means "to take away" or "to find how many are left."
A "minus" sign $\boxed{-}$ is the symbol for subtraction.

8 pieces **take away** 3 pieces = 5 pieces.

Fill in the blanks based on the picture above.

1. a)
$$\begin{array}{r} 8 \\ -\ 3 \\ \hline \boxed{} \end{array}$$

b) $8 - \boxed{} = \boxed{}$

2. a) If you have 5 circles and take away 2, how many circles are left? _____

b) $5 - 2 =$ _____

3. Draw 6 circles inside the rectangle. Cross out 3 circles.

How many circles are left?

$6 - \boxed{} = \boxed{}$

Practice Helps

Each column should be done in less than one minute.

1. $8 - 7 =$ ___	21. $4 - 4 =$ ___	41. $13 - 6 =$ ___	61. $13 - 9 =$ ___
2. $5 - 3 =$ ___	22. $8 - 0 =$ ___	42. $15 - 8 =$ ___	62. $12 - 3 =$ ___
3. $9 - 4 =$ ___	23. $6 - 5 =$ ___	43. $12 - 4 =$ ___	63. $11 - 3 =$ ___
4. $7 - 3 =$ ___	24. $9 - 9 =$ ___	44. $11 - 7 =$ ___	64. $12 - 6 =$ ___
5. $6 - 6 =$ ___	25. $7 - 4 =$ ___	45. $14 - 9 =$ ___	65. $11 - 2 =$ ___
6. $9 - 7 =$ ___	26. $4 - 2 =$ ___	46. $13 - 8 =$ ___	66. $12 - 7 =$ ___
7. $3 - 2 =$ ___	27. $1 - 1 =$ ___	47. $10 - 2 =$ ___	67. $15 - 5 =$ ___
8. $7 - 5 =$ ___	28. $3 - 1 =$ ___	48. $18 - 9 =$ ___	68. $10 - 3 =$ ___
9. $8 - 6 =$ ___	29. $7 - 7 =$ ___	49. $14 - 6 =$ ___	69. $16 - 7 =$ ___
10. $5 - 0 =$ ___	30. $4 - 1 =$ ___	50. $12 - 8 =$ ___	70. $11 - 1 =$ ___
11. $9 - 1 =$ ___	31. $1 - 0 =$ ___	51. $11 - 5 =$ ___	71. $14 - 3 =$ ___
12. $7 - 6 =$ ___	32. $8 - 2 =$ ___	52. $10 - 6 =$ ___	72. $10 - 5 =$ ___
13. $8 - 3 =$ ___	33. $9 - 3 =$ ___	53. $11 - 9 =$ ___	73. $12 - 9 =$ ___
14. $6 - 3 =$ ___	34. $2 - 1 =$ ___	54. $13 - 5 =$ ___	74. $11 - 4 =$ ___
15. $7 - 2 =$ ___	35. $3 - 0 =$ ___	55. $14 - 7 =$ ___	75. $15 - 6 =$ ___
16. $8 - 4 =$ ___	36. $6 - 4 =$ ___	56. $16 - 9 =$ ___	76. $10 - 7 =$ ___
17. $5 - 2 =$ ___	37. $7 - 1 =$ ___	57. $13 - 7 =$ ___	77. $12 - 2 =$ ___
18. $6 - 2 =$ ___	38. $9 - 2 =$ ___	58. $11 - 8 =$ ___	78. $14 - 4 =$ ___
19. $9 - 6 =$ ___	39. $9 - 5 =$ ___	59. $12 - 5 =$ ___	79. $11 - 6 =$ ___
20. $2 - 0 =$ ___	40. $6 - 1 =$ ___	60. $15 - 3 =$ ___	80. $14 - 8 =$ ___

Timed Subtraction Drill

Can you subtract these in less than 3 minutes?

1.	7 − 4	10.	8 − 7	19.	9 − 6	28.	10 − 5	37.	12 − 5	46.	11 − 3
2.	8 − 5	11.	9 − 5	20.	7 − 5	29.	9 − 8	38.	14 − 3	47.	12 − 8
3.	9 − 2	12.	4 − 4	21.	5 − 2	30.	10 − 7	39.	13 − 6	48.	6 − 4
4.	6 − 6	13.	10 − 2	22.	7 − 3	31.	12 − 9	40.	11 − 7	49.	14 − 6
5.	10 − 6	14.	7 − 6	23.	4 − 0	32.	14 − 4	41.	15 − 4	50.	12 − 2
6.	5 − 4	15.	9 − 3	24.	8 − 6	33.	13 − 8	42.	5 − 5	51.	4 − 3
7.	1 − 1	16.	6 − 5	25.	9 − 4	34.	11 − 5	43.	13 − 4	52.	13 − 7
8.	10 − 3	17.	8 − 8	26.	6 − 3	35.	15 − 7	44.	14 − 9	53.	8 − 0
9.	6 − 2	18.	5 − 3	27.	8 − 4	36.	9 − 9	45.	8 − 3	54.	12 − 6

Relating Addition and Subtraction

Addition and subtraction are opposites. Fill in the boxes.

1. If $6 + 4 = 10$, then $\boxed{10} - 4 = \boxed{}$

2. If $5 + 9 = 14$, then $\boxed{14} - 5 = \boxed{}$

3. If $3 + 4 = 7$, then $\boxed{} - 4 = \boxed{}$

4. If $5 + 1 = 6$, then $\boxed{} - 1 = \boxed{}$

5. If $7 + 0 = 7$, then $\boxed{} - 7 = \boxed{}$

6. If $8 + 5 = 13$, then $\boxed{} - 8 = \boxed{}$

7. $4 + \boxed{} = 11$, so $\boxed{} - 4 = \boxed{}$

8. $7 + \boxed{} = 10$, so $\boxed{} - 3 = \boxed{}$

9. $3 + \boxed{} = 12$, so $\boxed{} - 9 = \boxed{}$

10. $9 + \boxed{} = 9$, so $\boxed{} - 0 = \boxed{}$

11. $\boxed{} + 4 = 6$

12. $\boxed{} - 2 = 7$

13. $\boxed{} - 7 = 8$

14. $9 + \boxed{} = 15$

15. $\boxed{} + 6 = 14$

16. $\boxed{} - 8 = 16$

17. $11 + \boxed{} = 25$

18. $\boxed{} - 32 = 65$

19. $\boxed{} + 16 = 45$

20. $13 + \boxed{} = 21$

Subtracting Ones and Tens

STEP 1	STEP 2	STEP 3
Subtract the ones.	Subtract the tens.	To check subtraction, you must add.

STEP 1 — Subtract the ones.

$5 - 3 = 2$

$$\begin{array}{r} 7\,5 \\ -\ 2\,3 \\ \hline 2 \end{array}$$

STEP 2 — Subtract the tens.

$7 - 2 = 5$

$$\begin{array}{r} 7\,5 \\ -\ 2\,3 \\ \hline 5\,2 \end{array}$$

STEP 3 — To check subtraction, you must add.

These numbers must be equal.

check

$$\begin{array}{r} 2\,3 \\ +\ 5\,2 \\ \hline 7\,5 \end{array}$$

Subtract and check your answers.

1. a)

$$\begin{array}{r} 6\,8 \\ -\ 4\,1 \\ \hline \end{array} \rightarrow$$

check

b) If you get 68 when you check, the answer is correct. To check subtraction you must __ __ __ .

word

2.
$$\begin{array}{r} 9\,3 \\ -\ 6\,1 \\ \hline \end{array} \rightarrow$$
check

4.
$$\begin{array}{r} 5\,1 \\ -\ 2\,1 \\ \hline \end{array} \rightarrow$$
check

6.
$$\begin{array}{r} 6\,4 \\ -\ 5\,2 \\ \hline \end{array} \rightarrow$$
check

3.
$$\begin{array}{r} 3\,7 \\ -\ 1\,5 \\ \hline \end{array} \rightarrow$$
check

5.
$$\begin{array}{r} 6\,1 \\ -\ 1\,0 \\ \hline \end{array} \rightarrow$$
check

7.
$$\begin{array}{r} 7\,7 \\ -\ 5\,5 \\ \hline \end{array} \rightarrow$$
check

Think Zero

Sometimes you will need to think zero when subtracting.

• Subtract the ones.

• Subtract the tens.

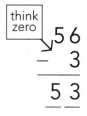

think zero

$$\begin{array}{r} 56 \\ -\ 3 \\ \hline 53 \end{array}$$

A.
$$\begin{array}{r} 38 \\ -\ 5 \\ \hline \underline{\ \ }3 \end{array}$$
finish

B.
$$\begin{array}{r} 75 \\ -53 \\ \hline \underline{\ \ }2 \end{array}$$
finish

C.
$$\begin{array}{r} 96 \\ -24 \\ \hline \underline{\ \ }2 \end{array}$$
finish

Subtract the numbers. Think zero where necessary.

1.
$$\begin{array}{r} 47 \\ -\ 5 \\ \hline \end{array}$$

5.
$$\begin{array}{r} 78 \\ -\ 6 \\ \hline \end{array}$$

9.
$$\begin{array}{r} 27 \\ -\ 3 \\ \hline \end{array}$$

13.
$$\begin{array}{r} 58 \\ -\ 2 \\ \hline \end{array}$$

2.
$$\begin{array}{r} 66 \\ -\ 4 \\ \hline \end{array}$$

6.
$$\begin{array}{r} 39 \\ -\ 8 \\ \hline \end{array}$$

10.
$$\begin{array}{r} 86 \\ -\ 5 \\ \hline \end{array}$$

14.
$$\begin{array}{r} 24 \\ -\ 4 \\ \hline \end{array}$$

3.
$$\begin{array}{r} 99 \\ -58 \\ \hline \end{array}$$

7.
$$\begin{array}{r} 87 \\ -67 \\ \hline \end{array}$$

11.
$$\begin{array}{r} 69 \\ -14 \\ \hline \end{array}$$

15.
$$\begin{array}{r} 53 \\ -21 \\ \hline \end{array}$$

4.
$$\begin{array}{r} 67 \\ -33 \\ \hline \end{array}$$

8.
$$\begin{array}{r} 95 \\ -40 \\ \hline \end{array}$$

12.
$$\begin{array}{r} 86 \\ -55 \\ \hline \end{array}$$

16.
$$\begin{array}{r} 78 \\ -37 \\ \hline \end{array}$$

Regrouping to Subtract

The number 25 can be written as 2 tens and 5 ones or as 1 ten and 15 ones.

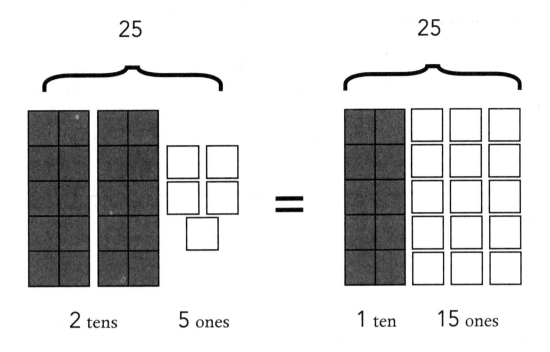

25	25
2 tens 5 ones	1 ten 15 ones

Regroup the numbers.

1. 43 = 4 tens and 3 ones = 3 tens and __1__ __3__ ones

2. 76 = 7 tens and 6 ones = 6 tens and ___ ___ ones

3. 92 = 9 tens and 2 ones = 8 tens and ___ ___ ones

4. 35 = 3 tens and 5 ones = 2 tens and ___ ___ ones

5. 18 = 1 ten and 8 ones = 0 tens and ___ ___ ones

6. 50 = 5 tens and 0 ones = 4 tens and ___ ___ ones

Subtraction Readiness

Sometimes we must regroup to subtract.

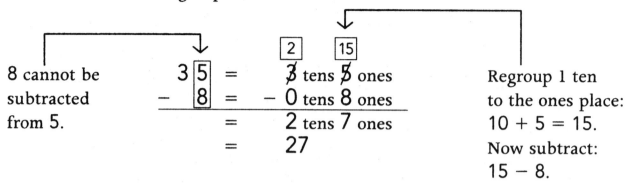

8 cannot be subtracted from 5.

$3\boxed{5}$ = $\boxed{2}$ ~~3~~ tens $\boxed{15}$ ~~5~~ ones
$-\ \boxed{8}$ = $-\ 0$ tens 8 ones
= $\ \ 2$ tens 7 ones
= $\ \ 27$

Regroup 1 ten to the ones place: $10 + 5 = 15$.
Now subtract: $15 - 8$.

Regroup and complete.

1.

83 = $\boxed{7}$ ~~8~~ tens $\boxed{13}$ ~~3~~ ones
$-\ 9$ = $-\ 0$ tens 9 ones
= $\underline{\ 7\ }$ tens $\underline{\ 4\ }$ ones
= $\underline{\hspace{1.5cm}}$

2.

64 = $\square$ ___ tens $\square$ ___ ones
$-\ 5$ = $-\ $ ___ tens ___ ones
= ___ tens ___ ones
= $\underline{\hspace{1.5cm}}$

3.

55 = $\square$ ___ tens $\square$ ___ ones
$-\ 7$ = $-\ $ ___ tens ___ ones
= ___ tens ___ ones
= $\underline{\hspace{1.5cm}}$

4.

57 = $\square$ ___ tens $\square$ ___ ones
$-\ 8$ = $-\ $ ___ tens ___ ones
= ___ tens ___ ones
= $\underline{\hspace{1.5cm}}$

5.

76 = $\square$ ___ tens $\square$ ___ ones
$-\ 7$ = $-\ $ ___ tens ___ ones
= ___ tens ___ ones
= $\underline{\hspace{1.5cm}}$

6.

62 = $\square$ ___ tens $\square$ ___ ones
$-\ 6$ = $-\ $ ___ tens ___ ones
= ___ tens ___ ones
= $\underline{\hspace{1.5cm}}$

Regrouping Tens to Ones

STEP 1

$$\begin{array}{r} 3|6 \\ -|7 \\ \hline \end{array}$$

The top digit 6 is smaller than the bottom digit 7, so you must regroup.

STEP 2

$$\begin{array}{r} \overset{2}{\cancel{3}}6 \\ -7 \\ \hline \end{array}$$

Regroup 1 ten, leaving 2 tens.

STEP 3

$$\begin{array}{r} \overset{2}{\cancel{3}}\overset{16}{\cancel{6}} \\ -7 \\ \hline \end{array}$$

You have regrouped 1 ten, so 10 + 6 = 16.

STEP 4

$$\begin{array}{r} \overset{2}{\cancel{3}}\overset{16}{\cancel{6}} \\ -7 \\ \hline \end{array}$$

finish

Now you can subtract. Start with the ones and work left.

Regroup and subtract.

1. $\begin{array}{r} \overset{1\,17}{\cancel{2}\cancel{7}} \\ -8 \\ \hline 9 \\ \end{array}$ finish

4. $\begin{array}{r} 93 \\ -5 \\ \hline \end{array}$

7. $\begin{array}{r} 51 \\ -4 \\ \hline \end{array}$

10. $\begin{array}{r} 35 \\ -6 \\ \hline \end{array}$

2. $\begin{array}{r} 22 \\ -4 \\ \hline \end{array}$

5. $\begin{array}{r} 64 \\ -7 \\ \hline \end{array}$

8. $\begin{array}{r} 67 \\ -9 \\ \hline \end{array}$

11. $\begin{array}{r} 53 \\ -8 \\ \hline \end{array}$

3. $\begin{array}{r} 87 \\ -8 \\ \hline \end{array}$

6. $\begin{array}{r} 62 \\ -4 \\ \hline \end{array}$

9. $\begin{array}{r} 30 \\ -3 \\ \hline \end{array}$

12. $\begin{array}{r} 77 \\ -9 \\ \hline \end{array}$

Regroup and Subtract

Subtract the numbers.

1.
$$
\begin{array}{r}
{}^{8}\!\!\not{9}\,{}^{13}\!\!\not{3} \\
-\ 5\ 9 \\
\hline
\end{array}
$$
finish

2.
$$
\begin{array}{r}
{}^{2}\!\!\not{3}\,{}^{18}\!\!\not{8} \\
-\ 1\ 9 \\
\hline
\end{array}
$$
finish

3.
$$
\begin{array}{r}
2\ 4 \\
-\ 1\ 5 \\
\hline
\end{array}
$$

4.
$$
\begin{array}{r}
9\ 7 \\
-\ 1\ 8 \\
\hline
\end{array}
$$

5.
$$
\begin{array}{r}
{}^{6}\!\!\not{7}\,{}^{10}\!\!\not{0} \\
-\ 3\ 4 \\
\hline
\end{array}
$$
finish

6.
$$
\begin{array}{r}
5\ 1 \\
-\ 2\ 2 \\
\hline
\end{array}
$$

7.
$$
\begin{array}{r}
6\ 0 \\
-\ 1\ 6 \\
\hline
\end{array}
$$

8.
$$
\begin{array}{r}
3\ 0 \\
-\ 2\ 4 \\
\hline
\end{array}
$$

9.
$$
\begin{array}{r}
{}^{3}\!\!\not{4}\,{}^{16}\!\!\not{6} \\
-\ 2\ 7 \\
\hline
\end{array}
$$
finish

10.
$$
\begin{array}{r}
7\ 6 \\
-\ 3\ 7 \\
\hline
\end{array}
$$

11.
$$
\begin{array}{r}
4\ 0 \\
-\ 1\ 3 \\
\hline
\end{array}
$$

12.
$$
\begin{array}{r}
8\ 4 \\
-\ 5\ 6 \\
\hline
\end{array}
$$

Whole Numbers: Addition & Subtraction

Deciding to Regroup

When you subtract, you must first decide whether or not to regroup.

$$\begin{array}{r} 3|3 \\ -2|9 \\ \hline \end{array}$$

Regroup?
(Yes) No   *(circle one)*
Why?
9 is larger than 3.

$$\begin{array}{r} 9|7 \\ -7|3 \\ \hline \end{array}$$

Regroup?
Yes (No)   *(circle one)*
Why?
3 is smaller than 7.

Decide whether to regroup or not. Give your reason. **Do not solve the problem.**

1.
$$\begin{array}{r} 97 \\ -48 \\ \hline \end{array}$$
Regroup?
Yes No   *(circle one)*
Why?

_____

4.
$$\begin{array}{r} 49 \\ -41 \\ \hline \end{array}$$
Regroup?
Yes No   *(circle one)*
Why?

_____

2.
$$\begin{array}{r} 56 \\ -37 \\ \hline \end{array}$$
Regroup?
Yes No   *(circle one)*
Why?

_____

5.
$$\begin{array}{r} 75 \\ -57 \\ \hline \end{array}$$
Regroup?
Yes No   *(circle one)*
Why?

_____

3.
$$\begin{array}{r} 49 \\ -14 \\ \hline \end{array}$$
Regroup?
Yes No   *(circle one)*
Why?

_____

6.
$$\begin{array}{r} 99 \\ -81 \\ \hline \end{array}$$
Regroup?
Yes No   *(circle one)*
Why?

_____

Regroup Only When Necessary

Regroup if necessary and subtract.

1.
$$\begin{array}{r} {}^{2\,13}\! 3\!\!\!\not{3} \\ -1\,9 \\ \hline \end{array}$$
finish

5.
$$\begin{array}{r} {}^{7\,10}\! 8\!\!\not{0} \\ -5\,6 \\ \hline \end{array}$$
finish

9.
$$\begin{array}{r} {}^{4\,14}\! 5\!\!\not{4} \\ -3\,9 \\ \hline \end{array}$$
finish

2.
$$\begin{array}{r} 8\,6 \\ -4\,2 \\ \hline 4 \end{array}$$
finish

6.
$$\begin{array}{r} 9\,8 \\ -9\,3 \\ \hline \end{array}$$

10.
$$\begin{array}{r} 6\,5 \\ -2\,5 \\ \hline \end{array}$$

3.
$$\begin{array}{r} 4\,3 \\ -\ \ 7 \\ \hline \end{array}$$

7.
$$\begin{array}{r} 6\,3 \\ -2\,4 \\ \hline \end{array}$$

11.
$$\begin{array}{r} 8\,7 \\ -2\,8 \\ \hline \end{array}$$

4.
$$\begin{array}{r} 7\,9 \\ -3\,5 \\ \hline \end{array}$$

8.
$$\begin{array}{r} 2\,9 \\ -\ \ 8 \\ \hline \end{array}$$

12.
$$\begin{array}{r} 3\,6 \\ -2\,0 \\ \hline \end{array}$$

Whole Numbers: Addition & Subtraction

Subtract and Check

To check subtraction you must add.

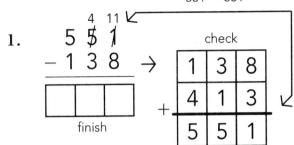

1.

$$551 = 551$$

```
   4 11
5 5̶ 1̶
-1 3 8
```
finish

check

1	3	8
4	1	3
5	5	1

+

2.
```
  9 4 2
- 7 1 3
```
check
+

3.
```
  6 5 0
- 4 2 3
```
check
+

4.
```
  3 6 6
- 2 3 7
```
check
+

5.
```
  1 8 4
-   1 5
```
check
+

6.
```
  9 4 3
- 5 0 4
```
check
+

7.
```
  2 3 4
-   2 5
```
check
+

8.
```
  6 3 2
- 4 1 3
```
check
+

Regrouping Tens and Hundreds

STEP 1

$$\begin{array}{r} 5\boxed{2}8 \\ -\ 5\boxed{3} \\ \hline \boxed{5} \end{array}$$

The top digit 2 is smaller than the bottom digit 5, so you must regroup.

STEP 2

$$\begin{array}{r} {}^{4}\cancel{5}28 \\ -\ \ 53 \\ \hline 5 \end{array}$$

Regroup 1 hundred, leaving 4 hundreds.

STEP 3

$$\begin{array}{r} {}^{4\ 12}\cancel{5}\cancel{2}8 \\ -\ \ 53 \\ \hline 5 \end{array}$$

1 hundred = 10 tens
10 + 2 = 12

STEP 4

$$\begin{array}{r} {}^{4\ 12}\cancel{5}\cancel{2}8 \\ -\ \ 53 \\ \hline 5 \end{array}$$
finish

Now you can subtract. Start with the ones and work left.

Subtract. Start at the ones place and work to the left.

1.
$$\begin{array}{r} {}^{6\ 11}7\cancel{7}4 \\ -\ \ 62 \\ \hline 2 \end{array}$$
finish

4.
$$\begin{array}{r} {}^{5\ 13}6\cancel{3}7 \\ -\ \ 55 \\ \hline \ \ \ \end{array}$$
finish

7.
$$\begin{array}{r} {}^{3\ 15}4\cancel{5}8 \\ -\ \ 84 \\ \hline \ \ \ \end{array}$$
finish

2.
$$\begin{array}{r} {}^{2\ 17}3\cancel{7}3 \\ -\ \ 82 \\ \hline \ \ \ \end{array}$$
finish

5.
$$\begin{array}{r} 778 \\ -\ \ 87 \\ \hline \end{array}$$

8.
$$\begin{array}{r} 267 \\ -\ \ 72 \\ \hline \end{array}$$

3.
$$\begin{array}{r} 246 \\ -\ \ 73 \\ \hline \end{array}$$

6.
$$\begin{array}{r} 336 \\ -\ \ 91 \\ \hline \end{array}$$

9.
$$\begin{array}{r} 727 \\ -\ \ 25 \\ \hline \end{array}$$

Regrouping More Than Once

We may need to regroup more than one time.

STEP 1

8 can't be subtracted from 7.

$$\begin{array}{r} 83\boxed{7} \\ -45\boxed{8} \\ \hline \boxed{} \end{array}$$

Regroup from the tens place.

$$\begin{array}{r} {\scriptstyle 2\ 17} \\ 8\cancel{3}\cancel{7} \\ -458 \\ \hline 9 \end{array}$$

STEP 2

5 can't be subtracted from 2.

$$\begin{array}{r} {\scriptstyle 2\ 17} \\ 8\boxed{\cancel{3}}\cancel{7} \\ -4\boxed{5}8 \\ \hline \boxed{9} \end{array}$$

Regroup again from the hundreds place.

$$\begin{array}{r} {\scriptstyle 12} \\ {\scriptstyle 7\ \cancel{7}\ 17} \\ \cancel{8}\cancel{3}\cancel{7} \\ -458 \\ \hline 379 \end{array}$$

Regroup and subtract.

1.
$$\begin{array}{r} {\scriptstyle 13} \\ {\scriptstyle 6\ \cancel{3}\ 11} \\ \cancel{7}\cancel{4}\cancel{1} \\ -358 \\ \hline 3 \end{array}$$
finish

4.
$$\begin{array}{r} 381 \\ -192 \\ \hline \end{array}$$

7.
$$\begin{array}{r} 390 \\ -297 \\ \hline \end{array}$$

10.
$$\begin{array}{r} 368 \\ -\ \ 89 \\ \hline \end{array}$$

2.
$$\begin{array}{r} {\scriptstyle 5\ 10} \\ 5\cancel{6}\cancel{0} \\ -173 \\ \hline 7 \end{array}$$
finish

5.
$$\begin{array}{r} 233 \\ -\ \ 64 \\ \hline \end{array}$$

8.
$$\begin{array}{r} 135 \\ -\ \ 57 \\ \hline \end{array}$$

11.
$$\begin{array}{r} 243 \\ -\ \ 66 \\ \hline \end{array}$$

3.
$$\begin{array}{r} 472 \\ -295 \\ \hline \end{array}$$

6.
$$\begin{array}{r} 930 \\ -461 \\ \hline \end{array}$$

9.
$$\begin{array}{r} 338 \\ -\ \ 79 \\ \hline \end{array}$$

12.
$$\begin{array}{r} 550 \\ -371 \\ \hline \end{array}$$

Regrouping from Zero

Sometimes when you try to regroup, the next digit is a zero. You must then **move left** until you reach a number that is **not zero.**

70 tens − 1 ten = 69 tens
Regroup 1 ten to 10 ones

$$\begin{array}{r} \boxed{70}0 \\ -365 \\ \hline \end{array} = \begin{array}{r} \overset{6\,9}{7\!\!\!/\,\emptyset} \text{ tens} \quad \overset{10}{\emptyset} \text{ ones} \\ -\quad 36 \text{ tens} \quad 5 \text{ ones} \\ \hline \quad\;\; 33 \text{ tens} \quad 5 \text{ ones} \end{array}$$

A.
$$\begin{array}{r} \overset{6\,9\,10}{\boxed{7}\!\!\!/\,\emptyset\,\emptyset} \\ -365 \\ \hline \underline{} \end{array}$$
finish

Regroup and subtract.

1.
$$\begin{array}{r} \overset{2\,9\,10}{\boxed{3}\emptyset\,\emptyset} \\ -169 \\ \hline \end{array}$$
finish

4.
$$\begin{array}{r} \overset{3\,9\,10}{\boxed{4}\emptyset\,\emptyset} \\ -\;\;30 \\ \hline \end{array}$$
finish

7.
$$\begin{array}{r} \overset{-\;-\;10}{\boxed{6}\emptyset\,\emptyset} \\ -357 \\ \hline \end{array}$$
finish

10.
$$\begin{array}{r} \overset{-\;-\;10}{\boxed{9}\emptyset\,\emptyset} \\ -\;\;49 \\ \hline \end{array}$$
finish

2.
$$\begin{array}{r} \overset{9\,10}{\boxed{1}\emptyset\,\emptyset} \\ -\;\;63 \\ \hline \end{array}$$
finish

5.
$$\begin{array}{r} 600 \\ -378 \\ \hline \end{array}$$

8.
$$\begin{array}{r} 500 \\ -\;\;38 \\ \hline \end{array}$$

11.
$$\begin{array}{r} 200 \\ -160 \\ \hline \end{array}$$

3.
$$\begin{array}{r} 200 \\ -\;\;16 \\ \hline \end{array}$$

6.
$$\begin{array}{r} 700 \\ -284 \\ \hline \end{array}$$

9.
$$\begin{array}{r} 300 \\ -\;\;96 \\ \hline \end{array}$$

12.
$$\begin{array}{r} 700 \\ -199 \\ \hline \end{array}$$

Whole Numbers: Addition & Subtraction

Larger Numbers

500 tens − 1 ten = 499 tens
Regroup 1 ten to 10 ones.

800 tens − 1 ten = 799 tens
Regroup 1 ten to 10 ones.
10 + 4 = 14 ones

A.
$$\begin{array}{r} \overset{4\ \ 9\ \ 9\ \ 10}{5,000} \\ -\ \ \ \ 675 \\ \hline \end{array}$$
—,— — —
finish

B.
$$\begin{array}{r} \overset{7\ \ 9\ \ 9\ \ 14}{8,004} \\ -\ \ \ \ 268 \\ \hline \end{array}$$
—,— — —
finish

Regroup and subtract.

1.
$$\begin{array}{r} \overset{3\ \ 9\ \ 9\ \ 10}{4,000} \\ -\ \ \ \ 935 \\ \hline \end{array}$$
—,— — —
finish

5.
$$\begin{array}{r} \overset{-\ \ -\ \ 10}{6,005} \\ -\ \ \ \ 632 \\ \hline \end{array}$$
—,— — —
finish

9.
$$\begin{array}{r} \overset{-\ \ -\ \ -\ \ 18}{7,008} \\ -\ \ \ \ 599 \\ \hline \end{array}$$
—,— — —
finish

2.
$$\begin{array}{r} \overset{-\ \ -\ \ -\ \ 10}{2,000} \\ -\ \ \ \ 246 \\ \hline \end{array}$$
—,— — —
finish

6.
$$\begin{array}{r} 3,006 \\ -1,852 \\ \hline \end{array}$$

10.
$$\begin{array}{r} \overset{3\ \ 9\ \ 12}{4,027} \\ -\ \ \ \ 743 \\ \hline \end{array}$$
—,— — —

3.
$$\begin{array}{r} 9,000 \\ -\ \ \ \ 462 \\ \hline \end{array}$$

7.
$$\begin{array}{r} 8,002 \\ -3,064 \\ \hline \end{array}$$

11.
$$\begin{array}{r} 3,041 \\ -\ \ \ \ 532 \\ \hline \end{array}$$

4.
$$\begin{array}{r} 6,000 \\ -\ \ \ \ 507 \\ \hline \end{array}$$

8.
$$\begin{array}{r} 5,004 \\ -\ \ \ \ 83 \\ \hline \end{array}$$

12.
$$\begin{array}{r} 8,012 \\ -3,064 \\ \hline \end{array}$$

Subtraction Practice

Subtract the numbers. Regroup when necessary.

1. 94 – 73 =

5. 396 – 285 =

2.
```
   61
 − 45
```

6.
```
   247
 − 188
```

3. 476 – 54 =

7. 5,897 – 4,635 =

4.
```
   517
 −  48
```

8.
```
   8,225
 − 1,508
```

Using a Grid

Just as you do when you add, you must line up place values to subtract.

84 – 6

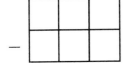

← You must subtract the smaller number.

A. Line up the numbers in the grid: 984 – 93.

← The smaller number (the number you subtract) goes here.

Line up place values **without working the problems.**

1. 45,154 – 342 =

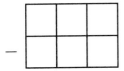

5. 21,673 – 1,842 =

2. From 645 subtract 15.

6. 8,915,056 – 125,604 =

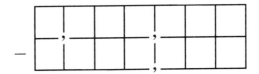

3. Take 884 from 1,995.

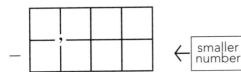

← smaller number

7. Subtract 340 from 2,019.

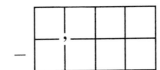

4. Find the difference between 68 and 7.

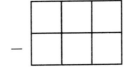

8. 235 minus 133 =

Line Up to Subtract

Line up and subtract. Regroup where necessary.

1. $295 - 52 =$

$$\begin{array}{r} 295 \\ -52 \\ \hline \rule{1.5cm}{0.4pt} \end{array}$$
finish

2. $530 - 68 =$

3. $400 - 45 =$

4. $600 - 185 =$

5. $4{,}953 - 405 =$

$$\begin{array}{r} {}^{4\ 13} \\ 4{,}9\cancel{5}\cancel{3} \\ -405 \\ \hline \rule{1.5cm}{0.4pt} \end{array}$$
finish

6. $3{,}000 - 650 =$

7. $4{,}276 - 92 =$

8. $4{,}091 - 103 =$

9. $2{,}859 - 199 =$

$$\begin{array}{r} {}^{7\ 15} \\ 2{,}8\cancel{5}9 \\ -199 \\ \hline \rule{1.5cm}{0.4pt} \end{array}$$
finish

10. $3{,}062 - 309 =$

11. $53{,}019 - 1{,}509 =$

12. $46{,}000 - 956 =$

Take Away

A. Take 35 away from 96.

B. Take 464 away from 1,493.

$$\begin{array}{r} 96 \\ -\ 35 \\ \hline \underline{\quad\quad} \end{array}$$
finish

The smaller number is the number you take away.

$$\begin{array}{r} {}^{8\ 13} \\ 1,49\cancel{3} \\ -\ \ 464 \\ \hline \underline{\text{—,———}} \end{array}$$
finish

Subtract the smaller number.

1. Take 19 away from 85.

4. Take 52 away from 341.

2. Take 37 away from 500.

5. Take 140 away from 2,057.

3. Take 38 away from 165.

6. Take 103 away from 750.

Find the Difference

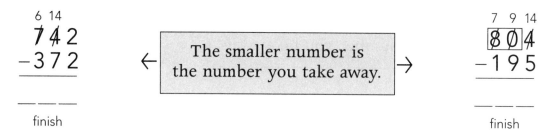

A. Find the difference between
742 and 372.

```
    6 14
   7 4 2
 - 3 7 2
 ———————
 finish
```

The smaller number is the number you take away.

B. Find the difference between
804 and 195.

```
   7 9 14
  8 0 4
- 1 9 5
———————
 finish
```

Subtract the smaller number.

1. Find the difference between 503 and 278.

4. Find the difference between 225 and 48.

2. Find the difference between 900 and 361.

5. Find the difference between 566 and 87.

3. Find the difference between 564 and 248.

6. Find the difference between 574 and 91.

Subtraction Review

1. If $6 + 7 = \boxed{}$, then $\boxed{} - 7 = 6$.

6.
$$\begin{array}{r} 837 \\ -458 \\ \hline \end{array}$$

2.
$$\begin{array}{r} 87 \\ -\ 4 \\ \hline \end{array}$$

7.
$$\begin{array}{r} 300 \\ -169 \\ \hline \end{array}$$

3.
$$\begin{array}{r} 479 \\ -316 \\ \hline \end{array}$$

8. $605 - 39 =$

4.
$$\begin{array}{r} 58 \\ -49 \\ \hline \end{array}$$
Regroup?
Yes   No   *(circle one)*
Why?
_____

9. $863 - 495 =$

5.
$$\begin{array}{r} 526 \\ -\ 69 \\ \hline \end{array}$$

10.
$$\begin{array}{r} 2,593 \\ -\ 764 \\ \hline \end{array}$$

Putting It All Together

NUMBER RELATION SYMBOLS		
SYMBOL	MEANING	EXAMPLES
=	is equal to	$13 + 3 = 8 + 8$ 16 is equal to 16
<	is less than	$6 - 2 < 5 + 0$ 4 is less than 5
>	is greater than	$9 > 6 + 2$ 9 is greater than 8

Place the symbol $<$, $=$, or $>$ in the ◯ to make each statement true.

1. $67 + 75$ ◯ $198 - 55$

2. $7 + 7 + 7 + 7$ ◯ 28

3. One hundred six ◯ 160

4. $7 + 5$ ◯ $19 - 6$

5. $295 - 52$ ◯ 240

6. $748 + 213$ ◯ 961

7. $831 - 205$ ◯ 626

8. $9,882$ ◯ $10,567 - 675$

9. 40 ◯ $8 + 8 + 8 + 8$

10. $93 - 59$ ◯ $17 + 17$

11. $50 - 22$ ◯ $9 + 9 + 9$

12. $9 + 8 + 7$ ◯ $7 + 9 + 8$

13. $500 - 49$ ◯ $116 + 335$

14. Four thousand eight ◯ $4,080$

Subtraction Problems

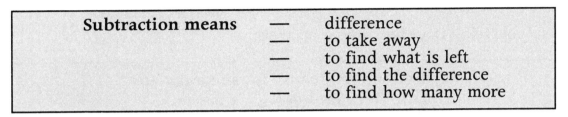

Subtraction means	—	difference
	—	to take away
	—	to find what is left
	—	to find the difference
	—	to find how many more

Complete each number sentence.

1. ◯ ◯ ◯ ⊗ ⊗

If you have 5 circles and take away 2, how many are left?

$\underline{\quad 5 \quad}$ $\underline{\quad - \quad}$ $\underline{\quad 2 \quad}$ = $\underline{\qquad}$
number · operation symbol · number · answer

4. ◯ ◯ ◯ ◯ ⊗ ⊗ ⊗

If you have 7 circles and take away 3 circles, how many are left?

$\underline{\qquad}$ $\underline{\qquad}$ $\underline{\qquad}$ = $\underline{\qquad}$
number · operation symbol · number · answer

2. Set A ◯ ◯ ◯ ◯ ◯

Set B ◯ ◯

There are 5 circles in Set A and 2 circles in Set B. Find the difference between the two sets.

$\underline{\qquad}$ $\underline{\qquad}$ $\underline{\qquad}$ = $\underline{\qquad}$
number · operation symbol · number · answer

5. Set A ◯ ◯ ◯

Set B ◯

There are 3 circles in Set A and one circle in Set B.

a) How many more circles are in A? $\underline{\quad}$

b) How many fewer circles are in B? $\underline{\quad}$

c) Compare the two sets with a number sentence.

$\underline{\qquad}$ $\underline{\qquad}$ $\underline{\qquad}$ = $\underline{\qquad}$
number · operation symbol · number · answer

3. Set A ◯ ◯ ◯ ◯ ◯ ◯

Set B ◯ ◯ ◯ ◯

If there are 6 circles in Set A and 4 circles in Set B, how many fewer circles are in Set B?

$\underline{\qquad}$ $\underline{\qquad}$ $\underline{\qquad}$ = $\underline{\qquad}$
number · operation symbol · number · answer

6. Set A has 10 circles.
Set B has 4 circles.

a) How many more circles are in Set A than in Set B? $\underline{\quad}$

b) How many fewer circles are in Set B than in Set A? $\underline{\quad}$

c) Compare the two sets.

$\underline{\qquad}$ $\underline{\qquad}$ $\underline{\qquad}$ = $\underline{\qquad}$
number · operation symbol · number · answer

Problem-Solving Strategies

1. Read over the problem several times to make sure you understand it.
2. Think about the facts in the problem and what you are being asked to find.
3. Complete the number sentence for each problem.
4. Ask yourself, "Does the answer make sense?"

> **Subtraction means "to take away" or**
> **"to find out how many are left."**
>
> Clue words can help you decide to subtract.
>
> | minus | less than | fewer | how many are not |
> |-------|-----------|-------|------------------|
> | take away | difference | how many are left | how many more than |

Complete each number sentence.

1. 7 minus 3 equals what number?

   _____ _____ _____   =   _____
   number operation number   answer
           symbol

2. 14 take away 8 is what number?

   _____ _____ _____   =   _____
   number operation number   answer
           symbol

3. What is 7 less than 15?

   _____ _____ _____   =   _____
   number operation number   answer
           symbol

4. Find the difference between 25 and 14.

   _____ _____ _____   =   _____
   number operation number   answer
           symbol

5. The flower shop has 95 roses. 36 are red roses. How many are not red roses?

   _____ _____ _____   =   _____
   number operation number   answer
           symbol

6. Neil saved $38. His sister saved only $10. How much more money did Neil save?

   _____ _____ _____   =   _____
   number operation number   answer
           symbol

7. A new radio costs $125. You have $75. How much more money do you need to buy the radio?

   _____ _____ _____   =   _____
   number operation number   answer
           symbol

8. Victor saved $25. He spent $12 for a new shirt. How much did he have left?

   _____ _____ _____   =   _____
   number operation number   answer
           symbol

Subtraction Word Problems

A number sentence can help you organize information and solve the problem.

Complete each number sentence.

1. Chris received 53 votes and Bobbi received 98 votes. How many more votes did Bobbi get than Chris?

    $\underline{98}$ $-$ $\underline{53}$ $=$ $\underline{}$

    number   operation   number   answer
    symbol

2. 84 letters were mailed on Wednesday and 36 on Friday. How many fewer letters were mailed on Friday than on Wednesday?

    $\underline{}$ $\underline{}$ $\underline{}$ $=$ $\underline{}$

    number   operation   number   answer
    symbol

3. Mr. Maxwell saved $125 and Mr. Kay saved $95. How much more money did Mr. Maxwell save than Mr. Kay?

    $\underline{}$ $\underline{}$ $\underline{}$ $=$ $\underline{}$

    number   operation   number   answer
    symbol

4. 35 students attended the party. 16 wanted ice cream. How many did not want ice cream?

    $\underline{}$ $\underline{}$ $\underline{}$ $=$ $\underline{}$

    number   operation   number   answer
    symbol

5. Roberta Landscaping planted 78 trees. 15 trees died. How many trees lived?

    $\underline{}$ $\underline{}$ $\underline{}$ $=$ $\underline{}$

    number   operation   number   answer
    symbol

6. The Lewchak family had $194. They spent $75 on food. How much money do they have left?

    $\underline{}$ $\underline{}$ $\underline{}$ $=$ $\underline{}$

    number   operation   number   answer
    symbol

7. The regular price of a television set is $545. The sale price is $470. How much more is the regular price than the sale price?

    $\underline{}$ $\underline{}$ $\underline{}$ $=$ $\underline{}$

    number   operation   number   answer
    symbol

8. Peter bought a suit on sale for $175. The regular price was $215. How much did he save by buying it on sale?

    $\underline{}$ $\underline{}$ $\underline{}$ $=$ $\underline{}$

    number   operation   number   answer
    symbol

9. Ted has $75. I have $15. How much more money does Ted have?

    $\underline{}$ $\underline{}$ $\underline{}$ $=$ $\underline{}$

    number   operation   number   answer
    symbol

10. Tara had $75 in her checking account. She wrote a check for $32. How much is left?

    $\underline{}$ $\underline{}$ $\underline{}$ $=$ $\underline{}$

    number   operation   number   answer
    symbol

Think About the Facts

Dorothy bought $77 worth of groceries.
She had $8 in coupons.

Question: <u>How much money</u>

<u>did she spend on groceries?</u>

<u>$77</u>	<u>—</u>	<u>$8</u>	=	<u>$69</u>
number	operation symbol	number		answer

Write a question and complete a number sentence for each problem. Choose to use addition or subtraction.

1. Adrian's math class has 39 students. His English class has 16 students.

Question: <u>How many students are</u>

<u>there in all?</u>

___	___	___	=	___
number	operation symbol	number		answer

2. Reedi's monthly car payment is $260. Her monthly rent payment is $195.

Question: _____

_____

___	___	___	=	___
number	operation symbol	number		answer

3. Will's monthly earnings are $1,537. His monthly bills come to $895.

Question: _____

_____

___	___	___	=	___
number	operation symbol	number		answer

4. Mary Lou drove 198 miles on Monday and 239 miles on Tuesday.

Question: _____

_____

___	___	___	=	___
number	operation symbol	number		answer

5. The theatre seats 497 people. 239 people attended the show.

Question: _____

_____

___	___	___	=	___
number	operation symbol	number		answer

6. Marg had 125 baseball cards. Sheena had 65 baseball cards.

Question: _____

_____

___	___	___	=	___
number	operation symbol	number		answer

Using Symbols

NUMBER RELATION SYMBOLS
$<$ less than
$>$ greater than
$\neq$ not equal to

Complete each number sentence to make it true.

1. Adam saved $15, and Kevin saved $21.
   Kevin saved how many more dollars than Adam?

   a) Adam saved $6 less than Kevin.

   $\underset{\text{Kevin}}{21} - \underset{\text{number}}{\underline{\hphantom{XXX}}} = \underset{\text{Adam}}{15}$

   b) Kevin saved $6 more than Adam.

   $\underset{\text{Adam}}{15} + \underset{\text{number}}{\underline{\hphantom{XXX}}} = \underset{\text{Kevin}}{21}$

   c) Kevin saved a greater amount
      of money than Adam.

   $\underset{\text{Kevin}}{\underline{\hphantom{XXX}}} > \underset{\text{Adam}}{\underline{\hphantom{XXX}}}$

   d) Adam saved a lesser amount of
      money than Kevin.

   $\underset{\text{Adam}}{\underline{\hphantom{XXX}}} < \underset{\text{Kevin}}{\underline{\hphantom{XXX}}}$

   e) Adam and Kevin did not save
      the same amount of money.

   $\underset{\text{Adam}}{15} \quad \underset{\text{symbol}}{\underline{\hphantom{XXX}}} \quad \underset{\text{Kevin}}{21}$

2. Jordan saved $16, and Steve saved $9.
   Jordan saved how many more dollars than Steve?

   a) Steve saved $7 less than Jordan.

   $\underset{\text{Jordan}}{16} - \underset{\text{number}}{\underline{\hphantom{XXX}}} = \underset{\text{Steve}}{9}$

   b) Jordan saved $7 more than Steve.

   $\underset{\text{Steve}}{9} + \underset{\text{number}}{\underline{\hphantom{XXX}}} = \underset{\text{Jordan}}{16}$

   c) Jordan saved a greater amount
      of money than Steve.

   $\underset{\text{Jordan}}{\underline{\hphantom{XXX}}} > \underset{\text{Steve}}{\underline{\hphantom{XXX}}}$

   d) Steve saved a lesser amount
      of money than Jordan.

   $\underset{\text{Steve}}{\underline{\hphantom{XXX}}} < \underset{\text{Jordan}}{\underline{\hphantom{XXX}}}$

   e) Jordan and Steve did not save
      the same amount of money.

   $\underset{\text{Jordan}}{16} \quad \underset{\text{symbol}}{\underline{\hphantom{XXX}}} \quad \underset{\text{Steve}}{9}$

Practice with Symbols

NUMBER RELATION SYMBOLS
< less than
> greater than
≠ not equal to

Complete each number sentence to make it true.

1. Bill spent $10. Joyce spent $3.
   How much more did Bill spend than Joyce ?

   a) Bill spent $7 more than Joyce.

   $$\underset{\text{Joyce}}{3} + \underset{\text{number}}{\underline{}} = \underset{\text{Bill}}{10}$$

   b) Joyce spent $7 less than Bill.

   $$\underset{\text{Bill}}{10} - \underset{\text{number}}{\underline{}} = \underset{\text{Joyce}}{3}$$

   c) Bill spent a greater amount of
      money than Joyce.

   $$\underset{\text{Bill}}{\underline{}} > \underset{\text{Joyce}}{\underline{}}$$

   d) Joyce spent a lesser amount
      of money than Bill.

   $$\underset{\text{Joyce}}{\underline{}} < \underset{\text{Bill}}{\underline{}}$$

   e) Bill and Joyce did not spend the
      same amount of money.

   $$\underset{\text{Bill}}{10} \quad \underset{\text{symbol}}{\underline{}} \quad \underset{\text{Joyce}}{3}$$

2. Jessica spent $35. Cassie spent $30.
   How much more did Jessica spend than Cassie?

   a) Jessica spent $5 more than Cassie.

   $$\underset{\text{Cassie}}{30} + \underset{\text{number}}{\underline{}} = \underset{\text{Jessica}}{35}$$

   b) Cassie spent $5 less than Jessica.

   $$\underset{\text{Jessica}}{35} - \underset{\text{number}}{\underline{}} = \underset{\text{Cassie}}{30}$$

   c) Jessica spent a greater amount of
      money than Cassie.

   $$\underset{\text{Jessica}}{\underline{}} > \underset{\text{Cassie}}{\underline{}}$$

   d) Cassie spent a lesser amount of
      money than Jessica.

   $$\underset{\text{Cassie}}{\underline{}} < \underset{\text{Jessica}}{\underline{}}$$

   e) Jessica and Cassie did not spend
      the same amount of money.

   $$\underset{\text{Jessica}}{35} \quad \underset{\text{symbol}}{\underline{}} \quad \underset{\text{Cassie}}{30}$$

Word Problem Review

Write a number sentence for each problem. Decide whether to use addition or subtraction.

1. In September 42 students registered for math and 57 registered for English. How many students registered altogether?

   42 ___ + ___ 57 ___ = ___
   number / operation symbol / number / answer

2. The price of your dinner came to $14. How much change will you get back from $20?

   ___ ___ ___ = ___
   number / operation symbol / number / answer

3. Abbey invited 28 people to her graduation party. 17 came to her party. How many people did not come to the party?

   ___ ___ ___ = ___
   number / operation symbol / number / answer

4. Mr. Brookes spent $9 for food and $15 for gasoline. How much did he spend altogether?

   ___ ___ ___ = ___
   number / operation symbol / number / answer

5. Ms. Maxwell saved $125 the first week and $250 the second week. What is Ms. Maxwell's combined savings for the two weeks?

   ___ ___ ___ = ___
   number / operation symbol / number / answer

6. 250 people came to see the first show. 138 people came to see the second show. Altogether how many people saw the two shows?

   ___ ___ ___ = ___
   number / operation symbol / number / answer

7. One cow weighed 704 pounds and another weighed 435 pounds. What is the combined weight of the two cows?

   ___ ___ ___ = ___
   number / operation symbol / number / answer

8. Cleo wants to plant 235 trees. 68 are white pine trees. How many are not white pine trees?

   ___ ___ ___ = ___
   number / operation symbol / number / answer

9. Last year Christopher weighed 245 pounds. He now weighs 190 pounds. How many pounds has Christopher lost?

   ___ ___ ___ = ___
   number / operation symbol / number / answer

10. Sabeen bought a basket with 136 apples. She gave 25 away. How many does she have left?

    ___ ___ ___ = ___
    number / operation symbol / number / answer

Picture Problems

Use the picture or chart to solve each problem.

ON SALE — TODAY ONLY

Regular Price
$175

Sale Price — $118

1. How much less is the sale price? _____

Sweater
$35

Coat
$98

2. What is the total price of the sweater and coat? _____

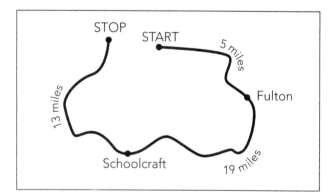

STOP

START

5 miles

Fulton

13 miles

Schoolcraft

19 miles

3. What is the total distance of the bike route? _____

TO (55) SOUTH

LAKEVIEW	128
CARSON CITY	34

4. How much farther is it to Lakeview than to Carson City? _____

MONEY SAVED	
Week	Savings
1	$35
2	$82
3	$27

5. How much was saved in all? _____

Use the chart below for questions 6 and 7.

Days	In	Out	Hours Worked
Monday	9:00 A.M.	5:00 P.M.	8
Tuesday	1:00 P.M.	4:00 P.M.	
Wednesday	7:00 A.M.	11:00 A.M.	
Thursday	10:00 A.M.	4:00 P.M.	
Friday	8:00 A.M.	1:00 P.M.	

6. Above is Pat's work schedule. How many hours did Pat work on
a) Monday? _____
b) Tuesday? _____
c) Wednesday? _____
d) Thursday? _____
e) Friday? _____

7. How many hours did Pat work that week? _____

Reading a Map

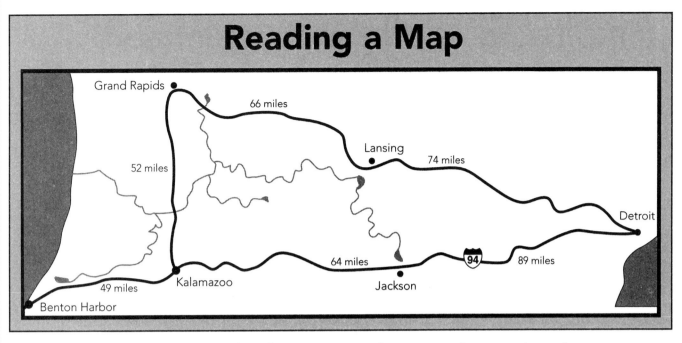

Grand Rapids
66 miles
52 miles
Lansing
74 miles
Detroit
64 miles
94
89 miles
Kalamazoo
Jackson
49 miles
Benton Harbor

Mr. Yuhas, a sporting goods salesman in Michigan, made stops in Kalamazoo, Grand Rapids, Lansing, Detroit, and Jackson.

1. Look at the map and fill in the distances.

    a) Benton Harbor to Kalamazoo: _____ miles

    b) Kalamazoo to Grand Rapids: _____ miles

    c) Grand Rapids to Lansing: _____ miles

    d) Lansing to Detroit: _____ miles

    e) Detroit to Jackson: _____ miles

    f) Jackson to Kalamazoo: _____ miles

    g) Kalamazoo to Benton Harbor: _____ miles

2. How many miles did Mr. Yuhas travel from Detroit to Benton Harbor on 94 ? _____

3. How many miles did he travel from Grand Rapids to Detroit? _____

4. How many miles did he travel from Detroit to Kalamazoo? _____

5. a) Is it farther from Grand Rapids to Detroit or Detroit to Kalamazoo?
    _____

    b) By how many miles? _____

6. Mr. Yuhas kept track of his sales for each city as follows:
    Kalamazoo $494, Grand Rapids $1,045, Lansing $697,
    Detroit $725, Jackson $530. What were his total sales? _____

Using Checks and Calendars

Mario Renda
124 E. Prairie
Natchez, New Mexico

1610

_____ , 20 ____

Pay to the order of   *Hills Drug Store* $49.00

Forty-nine and 00/100   DOLLARS

Second American Bank   *Mario Renda*

⑈03746⑈ 0311⑈ 0079125⑈

Use the check above to answer questions 1–4.

1. Fill in today's date in the shaded space provided for Mario on the above check.

2. If Mario had $245 in his checking account, how much money will be in his account after writing the $49 check? _____

3. If Mario writes another check for $158, how will he write this in words? _____ DOLLARS.

4. How much money will be left in his checking account after writing the $158 check? _____

JUNE						
S	M	T	W	T	F	S
				1	2	3
4	5	6	7	8	9	10
11	12	13	14	15	16	17
18	19	20	21	22	23	24
25	26	27	28	29	30	

Use the calendar to answer questions 5–8.

5. Is June 4 on a Saturday or Sunday? _____

6. Mario gets paid on June 2 and again every two weeks. Circle on the calendar Mario's paydays in June. What days are circled? _____

7. On June 5 the doctor told Mario to come back in three weeks. On what date will Mario go back to see the doctor? _____

8. Mario plans to leave for a vacation trip after work on June 9. He will return home June 23. How many days will he be on vacation? _____

Comparing Prices

Aaron wants to buy either a truck or a Jeep. After looking around, he was quoted the following prices:

TRUCK	
Original Price	$15,671
Sale Price	$14,398
OPTIONS	
Power Steering	$275
Road Wheels	$ 90
Air Conditioning	$900

JEEP	
Original Price	$18,567
Sale Price	$17,299
OPTIONS	
Radio AM/FM Stereo	$273
Rear Window Defroster	$ 96
Fold-Down Rear Seat	$395

1. a) How much could Aaron save by buying the truck on sale? _____

   b) By buying the Jeep on sale? _____

2. What is the total cost of the truck options? _____

3. What is the total cost of the Jeep options? _____

4. What is the difference between the cost of the truck options and the cost of the Jeep options? _____

5. a) Do you think Aaron will buy the truck or the Jeep? _____

   b) Why? _____

   _____

Life-Skills Math Review

Solve each problem. Write the answer on the line.

1. A video game is on sale at the local computer store for $25. The regular price is $33. How much more is the regular price?

    Answer: _____

2. The students at York High School sold 143 tickets for the Friday night concert, 122 tickets for the Saturday night concert, and 78 tickets for the Sunday night concert. How many tickets did they sell in all?

    Answer: _____

3. Jane is driving from Wheaton to downtown Chicago — a distance of 40 miles. She sees a sign that says "Downtown Chicago: 7 Miles." How far has she driven?

    Answer: _____

4. Julia has $832 in her checking account. She writes a check for $132 on Monday and another check for $538 on Tuesday. How much money is left in Julia's checking account?

    Answer: _____

5. Leah bought a car for $17,443. The options she chose totaled $1,985. How much would she have paid without the options?

    Answer: _____

Addition and Subtraction Review

Solve each problem. Write the answer on the line.

1. $83 + 347 + 254 =$

    Answer: _____

2. $2,398 - 1,543 =$

    Answer: _____

3. Laura has 35 flowers to sell at the fundraiser.
   By the end of the day she had 7 left. How many
   flowers did Laura sell?

    Answer: _____

4. Shayne spent $4 of his $12 allowance.
   How much money does he have left?

    Answer: _____

5. Pishka bought 144 flowers to plant in her garden.
   Her neighbor gave her 72 more. How many flowers
   does she have in all?

    Answer: _____

1.
$$34$$
$$+67$$

Answer: _____

2. A florist has 86 roses. 28 are red roses. How many are not red roses?

Answer: _____

3.
$$12,607$$
$$8,348$$
$$493$$
$$+ \ 5,726$$

Answer: _____

4. Henrick added $236 to his savings of $1,340. How much does he have in his savings altogether?

Answer: _____

5. $347 + 1,279 + 856 =$

Answer: _____

6. $7,100 - 489 =$

Answer: _____

7. $3,105 + 827 =$

Answer: _____

8.
$$7,016$$
$$- \ \ 307$$

Answer: _____

9. In 1990 the population of Richmond was 35,414. According to the census in 2000, the population was 3,837 more than in 1990. What was the population in 2000?

Answer: _____

10. $573 + 408 =$

Answer: _____

11. The Declaration of Independence was adopted by the Continental Congress in 1776. How old was the Declaration of Independence in 2002?

Answer: _____

12. Take 246 from 642.

Answer: _____

13. $4{,}258 + 67 + 2{,}308 =$

Answer: _____

14. From 834 subtract 86.

Answer: _____

15. Take 406 from 6,309.

Answer: _____

16. The Ohio River is 981 miles long, and the Mississippi River is 2,340 miles long. How much longer is the Mississippi River than the Ohio River?

Answer: _____

17. Meg's credit card purchases total $305. The finance charge is $15. How much is Meg's total bill?

Answer: _____

18. From 2,500 take 1,097.

Answer: _____

19. Heather makes $26,250 a year. If she gets a yearly raise of $1,375, what will be her new salary?

Answer: _____

20. Mereille wants to buy a bike that costs $205. She has saved $117. How much more money does Mereille need to save?

Answer: _____

Evaluation Chart

On the following chart, circle the number of any problem you missed. The column after the problem number tells you the pages where those problems are taught. You should review the sections for any problem you missed.

Skill Area	Posttest Problem Number	Skill Section	Review Page
Addition	1, 3, 5, 7, 10, 13	17–31	32
Subtraction	6, 8, 12, 14, 15, 18	39–60	61
Addition Word Problems	4, 9, 17, 19	33–37	38
Subtraction Word Problems	2, 11, 16, 20	63–68	69
Life-Skills Math	All	70–73	74

addition to combine numbers and find a total

$$
\begin{array}{r}
2 \\
+\ 2 \\
\hline
4
\end{array}
$$

attendance the number of people at an event

> Attendance at the football game was 450.

check to make sure that a problem has been completed correctly

checking account a bank account used to pay bills

> I pay my bills with checks from my checking account.

digit one of the ten numbers: 0, 1, 2, 3, 4, 5, 6, 7, 8, 9

difference the answer to a subtraction problem

$$5 - 3 = 2$$
↑

expanded form to write in long form

$$84 = 80 + 4$$

general admission the cost of attending an event (concert, play, etc.)

> Lisa said, "General admission to the concert is $15.00. Reserved seats cost more."

minus to subtract

number relation symbol symbols that compare two numbers

> For example:
>
> | | |
> |---|---|
> | less than | $<$ |
> | greater than | $>$ |
> | equal to | $=$ |
> | not equal to | $\neq$ |

15 is greater than 9 **OR** $15 > 9$

number sentence a complete math problem: $9 + 3 = 12$

operation symbol the symbol $(+, -, \times, \div)$ that tells you what to do with a math problem

$$16 + 4 = 20$$
↑
add

opposite to do the reverse (as in add and subtract)

$$5 + 3 = 8 \qquad 8 - 3 = 5$$

options choices

> The salesman explained, "There are several options available for the car including the type of stereo and the color of the seats."

pattern a consistent series of numbers: 2, 4, 6, 8, 10

place value   the name given to the space where a number is written

1,275
    ↑

The 2 is in the hundreds place.

plus   to add

quoted   a stated price

The car dealer quoted me a good price for the car I want to buy.

registered   to sign up for something such as a class

I registered for Pre-GED classes at the local adult learning center.

regroup (borrow)   to shift numbers to a lower place value

$$\begin{array}{r} \overset{2\ 12}{\cancel{3\,2}} \\ -\ 8 \\ \hline 24 \end{array}$$

regroup (carry)   to shift numbers to a higher place value

$$\begin{array}{r} \overset{1}{} \\ 18 \\ +\ 4 \\ \hline 22 \end{array}$$

regular price   the cost of an item not on sale

I paid regular price because the shirt was not on sale.

reserved seats   seats that are being held for someone

Lisa said, "The reserved seats cost $25.00."

sale price   the reduced cost of an item

The sale price of the shirt is very low.

subtraction   to find the difference between numbers

$$\begin{array}{r} 5 \\ -\ 3 \\ \hline 2 \end{array}$$

sum   the answer to an addition problem

symbol   a written sign used to represent an operation

$$2 + 2 = 4$$
    ↑   ↑
  add equals

take away   to subtract